Mercury and Methylmercury Toxicology and Risk Assessment

Mercury and Methylmercury Toxicology and Risk Assessment

Special Issue Editor

Laurie Hing Man Chan

MDPI • Basel • Beijing • Wuhan • Barcelona • Belgrade

MDPI

Special Issue Editor
Laurie Hing Man Chan
University of Ottawa
Canada

Editorial Office
MDPI
St. Alban-Anlage 66
4052 Basel, Switzerland

This is a reprint of articles from the Special Issue published online in the open access journal *Toxics* (ISSN 2305-6304) from 2018 to 2019 (available at: https://www.mdpi.com/journal/toxics/special_issues/Mercury_and_Methylmercury_Toxicology_and_Risk_Assessment)

For citation purposes, cite each article independently as indicated on the article page online and as indicated below:

LastName, A.A.; LastName, B.B.; LastName, C.C. Article Title. *Journal Name* **Year**, *Article Number*, Page Range.

ISBN 978-3-03897-970-8 (Pbk)
ISBN 978-3-03897-971-5 (PDF)

Cover image courtesy of Laurie Chan.

Contents

About the Special Issue Editor

Laurie Hing Man Chan is a professor and Canada Research Chair in Toxicology and Environmental Health at the University of Ottawa in Canada. Prof. Chan is a world-renowned expert in mercury toxicology and has published over 250 peer-reviewed scientific papers. He has served as an advisor for international and national governments and organizations and numerous indigenous communities on environmental health issues. Prof. Chan is a Fulbright Scholar and a Fellow of the Canadian Academy of Health Sciences.

toxics

MDPI

Editorial

Advances in Methylmercury Toxicology and Risk Assessment

Hing Man Chan

Department of Biology, University of Ottawa, 30 Marie Curie, Ottawa, ON K1N 6N5, Canada;
laurie.chan@uottawa.ca; Tel.: +613-562-5800 (ext. 7116)

Received: 19 March 2019; Accepted: 22 March 2019; Published: 1 April 2019

Mercury (Hg) is a global pollutant that affects the health of both humans and ecosystems. Wildlife and humans are exposed to Hg primarily through diet in the form of methylmercury (MeHg) because MeHg bioaccumulates and biomagnifies in the food web. Among Hg species, MeHg toxicity is significantly more damaging to the nervous system in early developmental stages because it leads to alterations in both structure and function. Even at low doses, prenatal exposure to MeHg can disrupt fetal brain development leading to long-term effects even in adulthood.

This Special Issue collected three review papers and six research articles that report the latest findings on the mechanisms of MeHg toxicology and its impacts on environmental health. Despite well-described neurobehavioral effects, the mechanisms of MeHg-induced toxicity are not completely understood. Antunes dos Santos et al. [1] reviewed the latest evidence indicating oxidative stress to be an important molecular mechanism in MeHg-induced intoxication. They have also highlighted the role of the PI3K/Akt signalling pathway and the nuclear transcription factor NF-E2-related factor 2 (Nrf2) in MeHg-induced redox imbalance. These results are important for future research in identifying the adverse outcome pathway for MeHg. In addition to its neurotoxicity, MeHg exposure has also been reported to affect heart health; however, evidence remains inconsistent. Karita et al. [2] reviewed the latest evidence on the association between MeHg and heart rate variability (HRV). They identified 13 studies examining the effect of MeHg exposure on HRV in human populations in the Faroe Islands, Seychelles, and other countries. They found that eight studies showed significant associations and five studies did not show any significant association. They concluded that the effects of MeHg might differ for prenatal and postnatal exposures and the HRV parameters calculated by frequency domain analysis were more sensitive than those by time domain analysis. These results suggest that HRV should be included in the risk characterization of MeHg. The third review by Sakamoto et al. [3] reported valuable lessons learned from the MeHg pollution in Minamata and the significance is discussed below.

Takahashi et al. [4] investigated the relationship between a chemokine, C-C motif Chemokine Ligand 4 (CCL4) expression and MeHg toxicity. Using both in vivo and in vitro approaches, they found that induction of CCL4 expression occurred prior to cytotoxicity caused by MeHg. They also showed that CCL4 is a protective factor against MeHg toxicity. Lee et al. [5] investigated the roles of a metal-binding protein metallothionein-III (MT-III) in chemokine gene expression changes in response to MeHg and Hg mercury vapour in the cerebrum and cerebellum in wild-type and MT-III knockout mice. They showed that the expression of Ccl12 and Cxcl12 was increased in the cerebrum by MeHg exposure in the wild-type mice only. Their results confirmed that MT-III plays a role in the expression of some chemokine genes in response to MeHg. The new understanding of these two studies on the protective roles of chemokines against the MeHg toxicity in the brain is important for the future development of treatments.

The mechanisms responsible for MeHg-induced changes in adult neuronal function, when exposure occurs primarily during fetal development, are not yet understood. Yuan et al. [6] dosed primary cortical precursor cells obtained from mice embryo with MeHg and found the cortical precursor exposed to extremely low dose (0.25 nM) of MeHg increased neuronal differentiation; while its proliferation was inhibited. In comparison, reduced neuronal differentiation was observed in the higher dose groups. These results suggest that sub-nanomolar MeHg exposure may deplete the pool of neural precursors by increasing premature neuronal differentiation. This can lead to long-term neurological effects in adulthood as neural precursors are important for adult neurogenesis. In contrast, the higher MeHg doses cause more immediate toxicity during infant development leading to neurobehaviour effects observed in infants and children. These results may explain the long delay effects of MeHg reported in Minamata, Japan as reported by Takaoka et al. [7]. In 2009, Takaoka et al. performed health surveys on 973 residents in the polluted area and 142 residents from a control area. Their results show that Minamata disease had spread outside of the central area and could still be observed almost 50 years after the Chisso Company's factory halted the dumping of Hg-polluted waste water in 1968. Results presented in this paper provide invaluable information on the long-term effects of MeHg, particularly on the wide range of clinical symptoms observed in adults who had no reported symptoms at younger age. Sakamoto et al. [3] reported a comprehensive review of the exposure and health effects which result from the MeHg pollution in Minamata. Information acquired during this review included the use of preserved umbilical cords to study the extent of pre-natal exposure, the higher sensitivity of male than female newborns to MeHg, and the neuropathology and kinetics of MeHg in fetuses. The sex-specific results were also confirmed by the results of a cohort study conducted by Tatsuta et al. [8]. To clarify the effects of MeHg on child development, a longitudinal prospective birth cohort study was conducted by the Tohoku Study of Child Development (TSCD) in Japan. Tatsuta et al. [8] reported the latest findings of the TSCD, showing that prenatal MeHg exposure affected psychomotor development in 18-month-olds, especially in males. The results of these studies provide important new information on the comparative use of different biomarkers, such as maternal blood or hair, for the dose-response assessment for MeHg on fetuses and child development.

Many toxic trace metals including Hg have been shown to induce immunotoxicity. However, there is a lack of specific biomarkers in peripheral blood samples for the effects of different chemical exposure. Monastero et al. [9] conducted a pilot study to correlate the blood levels of three metals, including cadmium, lead and mercury, with differences in the expression of 98 genes associated with stress, toxicity, inflammation, and autoimmunity in 24 participants from the Long Island Study of Seafood Consumption. They found that the expression of three genes (*IL1RAP*, *CXCR1*, and *ITGB2*) was negatively associated with Hg. This preliminary result suggests the possibility of using gene expression in blood samples to study the effects of MeHg on immune functions in future population studies.

In summary, this collection of papers on the toxicology and risk assessment provides useful new information on the mechanisms of MeHg toxicity and methods of improving risk assessment of MeHg.

References

1. Antunes dos Santos, A.; Ferrer, B.; Marques Gonçalves, F.; Tsatsakis, A.M.; Renieri, E.A.; Skalny, A.V.; Farina, M.; Rocha, J.B.T.; Aschner, M. Oxidative Stress in Methylmercury-Induced Cell Toxicity. *Toxics* **2018**, *6*, 47. [CrossRef] [PubMed]
2. Karita, K.; Iwata, T.; Maeda, E.; Sakamoto, M.; Murata, K. Assessment of Cardiac Autonomic Function in Relation to Methylmercury Neurotoxicity. *Toxics* **2018**, *6*, 38. [CrossRef] [PubMed]
3. Sakamoto, M.; Tatsuta, N.; Izumo, K.; Phan, P.T.; Vu, L.D.; Yamamoto, M.; Nakamura, M.; Nakai, K.; Murata, K. Health Impacts and Biomarkers of Prenatal Exposure to Methylmercury: Lessons from Minamata, Japan. *Toxics* **2018**, *6*, 45. [CrossRef] [PubMed]
4. Takahashi, T.; Kim, M.-S.; Iwai-Shimada, M.; Fujimura, M.; Toyama, T.; Naganuma, A.; Hwang, G.-W. Chemokine CCL4 Induced in Mouse Brain Has a Protective Role against Methylmercury Toxicity. *Toxics* **2018**, *6*, 36. [CrossRef] [PubMed]
5. Lee, J.-Y.; Tokumoto, M.; Hwang, G.-W.; Kim, M.-S.; Takahashi, T.; Naganuma, A.; Yoshida, M.; Satoh, M. Effect of Metallothionein-III on Mercury-Induced Chemokine Gene Expression. *Toxics* **2018**, *6*, 48. [CrossRef] [PubMed]

6. Yuan, X.; Wang, J.; Chan, H.M. Sub-Nanomolar Methylmercury Exposure Promotes Premature Differentiation of Murine Embryonic Neural Precursor at the Expense of Their Proliferation. *Toxics* **2018**, *6*, 61. [CrossRef] [PubMed]
7. Takaoka, S.; Fujino, T.; Kawakami, Y.; Shigeoka, S.-I.; Yorifuji, T. Survey of the Extent of the Persisting Effects of Methylmercury Pollution on the Inhabitants around the Shiranui Sea, Japan. *Toxics* **2018**, *6*, 39. [CrossRef] [PubMed]
8. Tatsuta, N.; Nakai, K.; Sakamoto, M.; Murata, K.; Satoh, H. Methylmercury Exposure and Developmental Outcomes in Tohoku Study of Child Development at 18 Months of Age. *Toxics* **2018**, *6*, 49. [CrossRef] [PubMed]
9. Monastero, R.N.; Vacchi-Suzzi, C.; Marsit, C.; Demple, B.; Meliker, J.R. Expression of Genes Involved in Stress, Toxicity, Inflammation, and Autoimmunity in Relation to Cadmium, Mercury, and Lead in Human Blood: A Pilot Study. *Toxics* **2018**, *6*, 35. [CrossRef] [PubMed]

Review

Oxidative Stress in Methylmercury-Induced Cell Toxicity

Alessandra Antunes dos Santos [1,*]**, Beatriz Ferrer** [1]**, Filipe Marques Gonçalves** [1]**,**
Aristides M. Tsatsakis [2]**, Elisavet A. Renieri** [2]**, Anatoly V. Skalny** [3,4,5]**, Marcelo Farina** [6]**,**
João B. T. Rocha [7] **and Michael Aschner** [1,*]

[1] Department of Molecular Pharmacology, Albert Einstein College of Medicine, Bronx, NY 10461, USA;
beatriz.ferrervillahoz@einstein.yu.edu (B.F.); filipe.goncalves@einstein.yu.edu (F.M.G.)
[2] Laboratory of Toxicology, Medical School, University of Crete, 71003 Heraklion, Greece;
aristsatsakis@gmail.com (A.M.T.); medp2011622@med.uoc.gr (E.A.R.)
[3] Department of Medical Elementology, Peoples' Friendship University of Russia (RUDN University),
Moscow 150000, Russia; skalny3@microelements.ru
[4] Laboratory of Biotechnology and Applied Bioelementology, Yaroslavl State University,
Yaroslavl 150014, Russia
[5] All-Russian Research Institute of Medicinal and Aromatic Plants (VILAR), Moscow 150000, Russia
[6] Department of Biochemistry, Federal University of Santa Catarina, Florianopolis 88040-900,
Santa Catarina, Brazil; marcelo.farina@ufsc.br
[7] Department of Biochemistry, Federal University of Santa Maria, Santa Maria 97105-900,
Rio Grande do Sul, Brazil; jbtrocha@yahoo.com.br
* Correspondence: Alessandra.antunes@einstein.yu.edu (A.A.d.S.); Michael.aschner@einstein.yu.edu (M.A.);
Tel.: +1-718-839-7920 (A.A.d.S.); +1-718-430-2317 (M.A.)

Received: 18 July 2018; Accepted: 7 August 2018; Published: 9 August 2018

Abstract: Methylmercury (MeHg) is a hazardous environmental pollutant, which elicits significant toxicity in humans. The accumulation of MeHg through the daily consumption of large predatory fish poses potential health risks, and the central nervous system (CNS) is the primary target of toxicity. Despite well-described neurobehavioral effects (i.e., motor impairment), the mechanisms of MeHg-induced toxicity are not completely understood. However, several lines of evidence point out the oxidative stress as an important molecular mechanism in MeHg-induced intoxication. Indeed, MeHg is a soft electrophile that preferentially interacts with nucleophilic groups (mainly thiols and selenols) from proteins and low-molecular-weight molecules. Such interaction contributes to the occurrence of oxidative stress, which can produce damage by several interacting mechanisms, impairing the function of various molecules (i.e., proteins, lipids, and nucleic acids), potentially resulting in modulation of different cellular signal transduction pathways. This review summarizes the general aspects regarding the interaction between MeHg with regulators of the antioxidant response system that are rich in thiol and selenol groups such as glutathione (GSH), and the selenoenzymes thioredoxin reductase (TrxR) and glutathione peroxidase (Gpx). A particular attention is directed towards the role of the PI3K/Akt signaling pathway and the nuclear transcription factor NF-E2-related factor 2 (Nrf2) in MeHg-induced redox imbalance.

Keywords: methylmercury; oxidative stress; molecular mechanisms

---

1. Introduction

Mercury (Hg) is a global pollutant ubiquitously present in the environment with a potential toxic effect in humans. Hg can be emitted to the atmosphere by natural sources (volcanoes and forest fire) or by anthropogenic sources (industrial activities, mining, and coal combustion), and deposits into aquatic systems, where it is primarily found [1]. The inorganic mercury in the aquatic environment can be biomethylated by aquatic sulfate-reducing bacteria forming methylmercury (MeHg) [2]. MeHg has a significant biomagnification potential and accumulates along the food chain by more than seven orders of magnitude, reaching higher concentrations in large predatory fish [3]. Therefore, fish consumption is the main source of MeHg exposure in humans [4,5]. After ingestion, MeHg is absorbed from the gastrointestinal tract (around 90–95%) and is distributed to all organs and systems, however, the central nervous system (CNS) is the most sensitive organ to MeHg-induced toxicity [6–11]. In this regard, it has been demonstrated that the developing CNS is particularly vulnerable to MeHg when compared to adults (i.e., the mature CNS) [12,13]. Several epidemiologic studies have shown that MeHg is able to produce severe cognitive deficits after prenatal and postnatal exposures [14–16]. Indeed, deficits in neurons and glia including abnormal migration, differentiation, and growth have been associated with prolonged pre- and/or perinatal exposure to MeHg, even at moderate doses [17,18]. In addition, studies with animals have shown the effects of developmental exposure to MeHg on animal behavior, such as reduced motor activity [19], decrease in memory [20–22] and learning [23], among others. Furthermore, there is evidence of a correlation between heavy metals' (including Hg) exposure and several skeletal deformities in fish larvae [24].

In addition, behavioral deficits in locomotor activity and motor performance were demonstrated in adult mice exposed to MeHg [25–27]. Indeed, the symptoms of MeHg poisoning in adults are frequently associated with loss of neuronal cells in the visual cortex and the cerebellum [28]. Even though the relationship between MeHg-induced motor deficit and cerebellar damage is a well-described phenomenon [29], the cellular mechanisms mediating MeHg-induced neurotoxicity have yet to be fully understood. In this regard, it is well documented that the electrophilic abilities of MeHg allow this toxicant to bind to soft nucleophilic groups, such as thiol (-SH) and selenol groups [1,12,30], disrupting the structure and activity of a large number of proteins and lead to disruption of various intracellular functions. Furthermore, a number of mechanisms have been identified as critical factors in MeHg-induced cell damage, including induction of oxidative stress via overproduction of reactive oxygen species (ROS) or reduction of antioxidant defenses and disruption of glutamate and calcium (Ca^{2+}) homeostasis [1,31–33]. The impairment of astrocytic glutamate transport by MeHg can lead to an overproduction of ROS, since increased glutamate concentrations in the synaptic cleft can cause hyperactivation of N-methyl D-aspartate (NMDA) type glutamate receptors, leading to an increase in intracellular Na^+ and Ca^{2+} [34], which is associated with generation of ROS [35]. In fact, inhibition of MeHg-induced ROS production was demonstrated in neurons treated with N-methyl-D-aspartate (NMDA) receptor antagonists [36]. Moreover, decreased antioxidant enzyme activity, such as glutathione peroxidase (Gpx), thioredoxin (Trx), and thioredoxin reductase (TrxR) [1,37–39], has been associated with increased ROS generation after exposures to MeHg, as well as the MeHg-induced mitochondrial dysfunction. MeHg can target specific thiol-containing proteins in the mitochondria, including respiratory chain complexes, leading to mitochondrial membrane potential loss and generation of ROS [40,41].

To summarize the current state of pertinent literature: (1) thiol and selenol groups have critical roles in governing MeHg pharmacokinetics; (2) the oxidative stress has a pivotal role in mediating MeHg-induced toxicity; (3) the nuclear factor erythroid 2–related factor 2 (Nrf2) can regulate the cellular response for attenuation of oxidative stress elicited by environmental toxicants [42]; (4) the phosphatidylinositol 3 (PI3) kinase/Akt pathway is essential for cell survival and can regulate Nrf2-mediated antioxidant and detoxification reactions [43,44]; this review summarizes the current knowledge regarding the molecular mechanisms involved in MeHg-induced oxidative stress and cell toxicity. Considerable attention has been directed towards the role of thiols and selenols in MeHg-induced redox imbalance, as well as the regulation/modulation induced by this metal of the transcription factor Nrf2 and PI3K/Akt signaling pathway.

2. Thiols and Selenols Play Fundamental Roles in MeHg-Induced Toxicity

Hg compounds react specifically with sulfhydryls (-SH) groups of cellular proteins and nonprotein molecules, forming stable complexes with defined stoichiometry, -S-Hg-R. MeHg's affinity for the anionic form of thiol (-SH) groups is extremely high and responsible for most of its toxicological effects [45]. Indeed, in biological media, MeHg is always complexed to -SH-containing ligands [46–50] and the therapeutic agents effective in reducing its body-burden contain -SH groups [51]. In this regard, it is noteworthy that the MeHg distribution to the different organs, as well as its excretion and, consequently, toxicity are likely related to its interaction with -SH groups on various biomolecules. MeHg interacts with the -SH group from non-protein molecules like glutathione (GSH), and also binds to L-cysteine forming the MeHg-L-cysteine complex, which is taken up by cells from different tissues by molecular mimicry as a surrogate of methionine [52]. Particularly important, MeHg-L-cysteine conjugates serve as a substrate for the neutral amino acid transporter, LAT-1, which transports MeHg (complexed with L-cysteine) across membranes [53]. Complexes of MeHg-cysteine and MeHg-GSH have been identified in blood [46,47], and complexes with MeHg-GSH in brain [48], liver [49], and bile [50].

Selenium (Se) is an essential trace element that is an integrant part of several enzymes (selenoenzymes) in the form of selenocysteine (Sec) [54]. Selenoenzymes play critical roles in the maintenance of cellular homeostasis since the reactivity of the selenol group (-SeH) in Sec is high [55,56]. Since selenols chemically resemble thiols (-SH) it also serves as a soft ligand for complexation with soft metals [57]. Given that under physiological conditions, selenols have a lower pKa (around 5.3) when compared to thiols, the fully ionized isoform (selenolates, -Se$^-$) are predominant and also more susceptible to electrophilic reactions by mercurials [58]. Five glutathione peroxidases (GPx) and three thioredoxin reductases (TrxR) are among the well-known redox-active selenoenzymes, with a selenol group in their redox active sites [55,56].

As previously mentioned, MeHg's affinity for sulfhydryls and selenols interferes with several important regulators of the antioxidant response that are rich in thiol and selenol groups. In this regard, the thioredoxin (Trx) and glutathione GSH systems play important roles in maintaining the redox balance in the brain, a tissue that is prone to oxidative stress due to its high-energy demand [59]. Taking it into account, the following sections address the predominant molecules targeted by MeHg in biological systems, as well as the effects of MeHg on the regulation/modulation of the PI3K/Akt signaling pathway and Nrf2 transcription factor

2.1. Glutathione (GSH)

GSH is the most abundant low molecular weight thiol compound synthesized in cells, reaching the concentrations of 1–10 mM, and it is the major antioxidant and redox buffer in human cells. In fact, GSH serves as a reducing agent for ROS and other unstable molecules, in the reaction catalyzed by GPx [60]. Several aspects of MeHg-induced neurotoxicity have been ascribed to GSH depletion. MeHg is able to interact with GSH, leading to the formation of an excretable GS–MeHg complex [61]. This interaction decreases the levels of GSH and, consequently, the GSH:GSSG (disulfide-oxidized) ratio, which contributes to the occurrence of oxidative stress [33]. It is worth mentioning that the pKa of GSH is high at physiological pH conditions and little GSH is ionized. However, at physiological pH values, glutathione S-transferase (GST) effectively lowers the pKa of the cysteine thiol of GSH from ~9.3 in solution to 6.5–7.4 at the active site of various GSTs, resulting in formation of the catalytically active nucleophilic thiolate anion (GS$^-$) [62,63].

MeHg has been shown to bind GSH, both in animal models and cell culture systems [64,65]. Indeed, several in vitro [66,67] and in vivo [12,68] evidences have shown that MeHg exposure causes GSH depletion. In addition, GSH and enzymes related to its synthesis are known important targets of MeHg-induced developmental neurotoxicity. For example, Stringari et al. have shown that prenatal exposure to MeHg disrupts the postnatal development of the GSH antioxidant system (GSH levels, GPx and glutathione reductase (GR) activities) in mouse brain. Furthermore, the authors demonstrated that the biochemical alterations endured even when mercury tissue levels decreased and became indistinguishable from those noted in pups born to control dams [68].

Epidemiological and animal studies have shown that MeHg causes loss of neuronal cells in specific brain regions including the visual cortex and the cerebellum [28], with massive loss of cerebellar granule cells (CGC) [69]. Interestingly, the increased sensitivity of CGC has been attributed to the relatively low GSH content of these cells, although it is not the only factor recognized [70–72]. In this regard, decreased GSH levels have been reported in the cerebral and cerebellar cortices of MeHg-exposed animals whose cortical mercury levels were in the low micromolar range [12,68]. In the mammalian cerebrum and cerebellum, the intracellular GSH concentrations are in the millimolar (mM) range, therefore the generation of ROS induced by MeHg might be related to GSH-independent mechanisms as well, in addition to its direct interaction with GSH [66,73].

GSH is synthesized by the sequential addition of cysteine to glutamate followed by the addition of glycine. In this sense, MeHg selectively inhibits astrocytic uptake systems for cystine and cysteine transport [74,75], compromising GSH synthesis and the CNS redox potential [76]. The MeHg-induced inhibition of cystine transport and astrocytic GSH production would ultimately lead to decreased neuronal GSH levels and increased glutamate toxicity. It is worth mentioning that one of the major mechanisms involved in MeHg-induced neurotoxicity is glutamate dyshomeostasis. Indeed, inhibition of vesicular glutamate uptake, increase in spontaneous release of glutamate from presynaptic terminals, and inhibition of glutamate uptake by astrocytes are crucial phenomena related to MeHg-mediated neurotoxicity [31,77,78].

Furthermore, studies on the toxicological relevance of MeHg–GSH interaction have demonstrated that strategies able to raise GSH levels are protective against MeHg-induced neurotoxicity [79,80]. For instance, increased ROS formation and depleted mono- and disulfide GSH were observed in neuronal, glial, and mixed cultures, and supplementation with exogenous GSH protected against the MeHg-induced neuronal death [81].

2.2. Selenoenzymes

Selenium plays a crucial role in antioxidant defense, as one Se atom is absolutely required at the active site of all selenoenzymes, such as GPx and TrxR, in the form of selenocystein [82]. GPx is an antioxidant enzyme that, in the presence of tripeptide GSH, adds two electrons to reduce H_2O_2 and lipid peroxides to water and lipid alcohols, respectively, while simultaneously oxidizing GSH to glutathione disulfide. The GPx/GSH system is thought to be a major defense in low-level oxidative stress, and decreased GPx activity or GHS levels may lead to the absence of adequate H_2O_2 and lipid peroxides detoxification, which may be converted to OH-radicals and lipid peroxyl radicals, respectively, by transition metals (Fe^{2+}) [83]. Thioredoxin (Trx) is essential for maintaining intracellular redox status. The expression of this small (12 kDa) ubiquitous thiol-active protein is induced by ROS and an elevated serum level may indicate a state of oxidative stress. In this regard, TrxR, a NADPH-dependent lipid hydroperoxide reductase, uses NADPH to maintain the levels of reduced Trx via a mechanism similar to that used by GR to maintain GSH levels, contributing to the maintenance of thiol redox homeostasis in proteins. Importantly, the inhibition of TrxR impairs the cyclical regeneration of Trx activity, as Trx remains in the oxidized state [84,85].

The activity of selenoenzymes, such as GPx, may be negatively affected by MeHg [38,41]. In addition, TxrR may be particularly sensitive to mercury compounds after both in vitro [86,87] and in vivo exposure [39,88]. Glutathione peroxidase isoform 1 (GPx1) is an initial molecular target in CGC exposed to nanomolar concentrations of MeHg (300 nM) for 4 days in vitro [37]. Interestingly, the authors have shown that GPx-1 inhibition occurred before any changes on potential targets that are normally affected by high-dose MeHg exposures. Moreover, Gpx1 overexpression was able to prevent MeHg-induced neuronal death in these cells [37], suggesting that selenoproteins may help to prevent oxidative stress induced when MeHg directly disrupts proteins involved in cellular redox system pathways. Of particular toxicological significance, the results obtained by Farina et al. 2009 [37] indicate the potential role of Gpx1 as a primary target of MeHg. In agreement, another study has demonstrated that MeHg decreased GPx activity in three different models: (a) in mouse brain after treatment with MeHg (40 mg/L in drinking water), (b) in mouse brain mitochondrial-enriched fractions isolated from MeHg-treated animals, and (c) in cultured human neuroblastoma SH-SY5Y cells [38]. The authors have concluded that the inhibition of this antioxidant enzyme may cause a significant increase of MeHg-induced impairment of cell viability and oxidative stress. Ruszkiewicz et al., 2016 have demonstrated that MeHg affects the antioxidant Trx and GSH systems in a sex-and structure-specific manner, and that these changes are not associated with altered mRNA expression, but rather posttranscriptional mechanisms [89]. Although the direct interaction of MeHg with the selenol group of the selenoproteins is an important molecular mechanism involved in the MeHg-induced cytotoxicity and decreased protein function [37,90], posttranscriptional events concerning mercury–selenium interaction seem to be also involved in the decreased activity of these selenoproteins [91,92].

Se deficiency has been reported to exacerbate the neurodevelopmental effects induced by MeHg [93,94]. In fact, maternal exposure to MeHg decreased Se concentration and impaired GPx activity in the neural tissue of the offspring, but not in the maternal neural tissue. In addition, MeHg exposure of Se-deficient perinatal mice resulted in retarded neurobehavioral development [94]. A relatively recent study corroborates the hypothesis that Se deficiency is a contributing factor to MeHg-induced toxicity. The authors have shown that selenoprotein genes from antioxidant pathways, such as members of the GPx and TrxR families, were exclusively downregulated by MeHg in whole zebrafish (Danio rerio) embryos, and then rescued by elevated Se levels [91]. These results suggest that Se can reduce the toxicity of MeHg, although the mechanisms behind this interaction have yet to be fully clarified. In addition, it has been shown that the selenocompounds diphenyl diselenide (PhSe)$_2$ and ebselen possess thiol peroxidase-like activity, and neuroprotective properties against MeHg-induced neurotoxicity in vivo and in vitro [95–98]. Using an in vitro experimental model of cultured human neuroblastoma cells (SH-SY5Y), a recent study has shown that MeHg was able to inhibit the activity of Gpx and TrxR, while coexposure to (PhSe)$_2$ and MeHg showed a protective effect on both the activity and expression of TrxR [99]. The authors have speculated that a direct interaction of MeHg with the selenol groups of these enzymes and/or a reduction in enzyme synthesis may possibly be related to the inhibition of TrxR and GPx activities [33].

3. MeHg and PI3K/Akt Signaling Pathway

The phosphoinositide-3 kinase (PI3K)/Akt signaling pathway has been extensively reviewed by several authors [100–102]. Briefly, activation of receptor tyrosine kinases (RTK) or G-protein-coupled receptors (GPCR) lead the recruitment of PI3K in the plasma membrane. Following its recruitment, PI3K is activated and phosphorylates phosphatidylinositol (4,5) biphosphate (PIP$_2$) to phosphatidylinositol (3,4 5)-triphosphate (PIP$_3$). PIP$_3$ recruits Akt to the plasmatic membrane, allowing its activation through phosphorylation on threonine 308 and serine 473 by phospho-inositide-dependent kinase-1 (PDK1) [103] and mammalian target of rapamycin complex 2 (mTORC2) [104], respectively. In addition, negative regulation of the pathway is mediated by phosphatase and tensin homolog (PTEN), which dephosphorylates PIP$_3$ to PIP$_2$ [105]. It is well known that the PI3K/Akt signaling pathway plays a key role in multiple cellular processes, including metabolism, cell survival, proliferation, and motility [102], to name a few. In this regard, the activation of the PI3K/Akt signaling pathway can culminate in the inhibition of apoptosis in several cellular models [106–108], whereas the inhibition of this pathway could be associated with decreased cell viability [109]. Furthermore, it has been shown that ROS can activate the PI3K/Akt signaling pathway and inhibit PTEN [110,111], thus, inducing the Akt activation, as shown in several cultured cell models, such as fibroblasts [112], vascular smooth muscle cells [113], and mesanglial cells [114]. Sonoda et al., 1999 have demonstrated that the activation of Akt by hydrogen peroxide (H$_2$O$_2$) in a glioblastoma cell line was blocked by wortmannin, a PI3K inhibitor, culminating in increased cell death [115].

Moreover, MeHg can also modulate the PI3K/Akt signaling pathway. Indeed, exposure to 100 nM MeHg for 24 h induced apoptosis in differentiating PC12 cells, and decreased Akt phosphorylation [116]. Corroborating these findings, Pierozan et al., 2017 have also demonstrated that MeHg reduced neuronal viability and induced caspase 3-dependent apoptosis with downregulated PI3K/Akt pathway in primary cortical neurons treated with 1 µM MeHg for 24 h. In this study, the authors noted increased oxidative damage, suggesting that MeHg downregulated Akt phosphorylation despite the noted increase in ROS production [117]. A recent in vivo study using 30-day-old pups from pregnant rats treated with 2 mg/kg MeHg from gestational day 5 until parturition, has shown that the pups prenatally exposed to MeHg failed to increase hippocampal oxidative stress, while the levels of Akt phosphorylation were decreased. These results suggest that MeHg may regulate the Akt signaling pathway by a mechanism independent of ROS generation [9].

On the other hand, a submicromolar MeHg exposure (1 µmol/L) induced Akt activation, probably secondary to increased ROS production (in pancreatic β cell—derived HIT-T15 cells). Moreover, the antioxidant *N*-acetyl-L-cysteine (NAC) prevented MeHg-induced upregulation of Akt phosphorylation but did not reverse PI3K activity in these cells [118]. Similar results were obtained in isolated mouse pancreatic islets [118], suggesting that MeHg-induced oxidative stress may be involved in the regulation of Akt phosphorylation independently of PI3K in pancreatic cells. The same authors have shown similar results in mice treated for 2–4 weeks with 20 µg/kg MeHg, demonstrating an increase in oxidative stress parameters and upregulation of Akt phosphorylation in pancreatic islets [118]. Furthermore, a low-dose MeHg exposure (up to 2 µM) increased Akt phosphorylation, presumably by inactivation of PTEN through S-mercuration (in neuroblastoma SH-SY5Y cell line). Pretreatment with the PI3K/Akt inhibitor, wortmannin, enhanced MeHg-induced cytotoxicity [119]. In this study, the authors have suggested that low-dose MeHg exposures might be associated with activation of cell survival responses [120].

4. MeHg Regulation of Nrf2 Activity

As mentioned before, the increase in ROS generation and the disruption of the antioxidant defense system are primary mechanisms related to MeHg-induced cell toxicity and neurodegeneration [1,33]. In this sense, a recent body of evidence has suggested that MeHg can modulate the Nrf2, the transcription factor involved in the regulation of the antioxidant defense system. Nrf2 is a transcription factor that belongs to cap 'n' collar (CNC) family, with a basic leucine zipper in its structure. This protein is an important regulator of the antioxidant cell response, regulating the expression of around 1055 genes involved not only in the antioxidant defenses, but also in cell proliferation, metabolism, and immune and detoxifying responses. Once activated, Nrf2 is translocated to the cell nucleus and forms heterodimers with other transcript such as c-Jun and small Maf proteins, binding to the antioxidant response element (ARE: 5′-GTGACNNNGC-3′), favoring the transcription of Nfr2-related genes [121–123]. The main regulator of Nrf2 activity is the Kelch-like ECH-associated protein 1 (Keap1) protein [123]. In basal conditions, Keap-1 is associated to Nrf2 inducing Nrf2 ubiquitination by E3 ubiquitin ligase, and consequent degradation through 26S proteasome. The Keap-1 structure contain some reactive cysteine residues (Cys151, Cys273, Cys288 and Cys297), which can be modified under oxidative stress conditions or by electrophilic compounds, promoting the disruption of the Keap-1/Nrf2 association and Nrf2 translocation to the cell nucleus, and consequent gene expression [121,123,124]. In addition, besides the negative regulation of Keap-1 on Nrf2 activity, the modulation of this protein by other signaling pathways has also been described. In fact, nuclear translocation of Nrf2 may involve the PKC and AMPK-dependent phosphorylation [125,126]. An association between PI3K/Akt activation and Nrf2 up-regulation has also been reported [127–130]. Once activated, Akt promotes the inhibition of GSK-3β through phosphorylation on ser-9. GSK-3β can induce Nrf2 phosphorylation, creating a degradation domain that is recognized by the ubiquitin ligase, targeting Nrf2 for proteasomal degradation [131–134]. GSK-3β can also phosphorylate a member of the src-kinase family named Fyn, which can phosphorylate Nrf2, favoring its export from the cell nucleus [130,135].

Since MeHg is a potent oxidative stress inducer and an electrophilic agent, the activation and up-regulation of Nrf2 have been described upon exposure to this metal. This hypothesis is supported by numerous data showing an increase in the Nrf2-related gene expression (such as: ho-1, nqo-1, cglc and Nrf2), and also an increase in the Nrf2 nuclear translocation, after exposures to MeHg in vitro [67,71,120,136] and in vivo [137]. Furthermore, it has been demonstrated that Nrf2 genetic knockdown can increase the susceptibility to MeHg toxicity, and treatment with compounds that are able to increase Nrf2 activity ameliorate some consequences of MeHg-induced toxicity [136,138]. In this sense, it seems that MeHg-induced Nrf2 activation is associated with a cellular defense mechanism in response to this metal exposure.

MeHg is able to disrupt the Nrf2/Keap-1 association, which may be associated with an increase in the Nrf2 activation [136,139]. It has been shown a direct interaction between MeHg and the cysteine residues in Keap-1 structure (Cys151, Cys368, and Cys489) [136,139,140]. As mentioned previously, MeHg can react with GSH to form a MeHg-SG adduct that is transferred to extracellular space by the multidrug-resistance associated protein (MRP) [1]. This GSH adduct readily undergoes S-transmercuration with cellular proteins, inducing the mercuration of Cys319 residue in the Keap-1 structure [141]. Cys151 residue is essential to electrophile-mediated disassociation of Keap1 from Nrf2, and The Cys319 residue is critical to the ubiquitin E3 ligase activity and consequent Nrf2 degradation [136,142,143]. In this sense, it is possible to speculate that the MeHg-induced modifications in these cysteine residues, through direct interaction with Cys151 and Cys319 or by ROS-mediated modifications, could disrupt the inhibitory effect of Keap-1 in Nrf2 activation, allowing its translocation to the cell nucleus. It is noteworthy that the cysteine residues in Keap-1 are also potential targets for oxidation mediated by ROS [124,130]. However, further studies are necessary to elucidate the role of ROS in MeHg-induced Nrf2 activation.

MeHg can induce the modulation of PI3k/Akt, and the inhibition of this signaling pathway was able to attenuate the MeHg-induced Nrf2 activation [71]. An increase in GSK-3β phosphorylation was also observed 6 h after an in vitro MeHg exposure in primary cortical astrocytes. MeHg decreased the expression of Fyn and Sp1 (the transcription factor associated to fyn expression), and Fyn nuclear localization, which may possibly suggest a Fyn downregulation in these cells [120]. Thus, it is speculated that MeHg promotes Akt activation and consequent inhibitory phosphorylation of GSK-3β. In these conditions, the downregulation of Fyn phosphorylation may inhibit the Fyn-mediated Nrf2 nuclear export, enhancing the Nrf2 nuclear localization. These authors have also shown that the increase in the Nrf2-related genes was followed by downregulation of Sp1-related genes (such as *fyn* and *tgf-β1*) [120]. This finding corroborates the idea that Sp1 can interact with Nrf2 at promoter sequences, suppressing the expression of Sp1 specific target genes [144].

5. Conclusions

MeHg is a hazardous environmental pollutant of great concern to public health because of its neurotoxic effects. Due to its electrophilic nature, MeHg can react with nucleophiles such as sulfhydryl- and selenol-containing proteins and low-molecular-weight molecules in biological systems. MeHg's affinity for the anionic form of thiol and selenol groups is extremely high and responsible for most of its toxicological effects. Indeed, MeHg can interfere with several crucial regulators of the antioxidant response that are rich in thiol and selenol groups such as glutathione (GSH), the major antioxidant and redox buffer in human cells, and the antioxidant selenoenzymes thioredoxin reductase (TrxR) and glutathione peroxidase (Gpx). Consequently, the excessive generation of ROS is an important phenomena that culminates in cell death. In this regard, antioxidant molecules have been reported as important protective agents against MeHg-induced toxicity. Our understanding of the critical targets of MeHg is incomplete, and detailed experimental data regarding the mechanism of action are needed. For instance, given that MeHg is known to generate ROS, and that mammalian cells activate Nrf2-mediated transcription in response to ROS, it is not surprising that Nrf2 activation has been demonstrated in response to MeHg exposure. However, relatively little information has been found about the potential mechanisms involved in MeHg-induced Nrf2 activation. In addition, the relationship between the PI3K/Akt signaling pathway and MeHg toxicity is still very limited, and consideration should be given to future research. In conclusion, the data presented in this review suggest that multiple mechanisms are involved in MeHg-induced oxidative stress and cell toxicity. Taken together, the data highlight several promising directions for future research in this area.

Funding: This research was funded by National Institute of Environmental Health Sciences, NIEHS R01ES07331, NIEHS R01ES10563 and NIEHS R01ES020852.

Conflicts of Interest: The authors declare no conflict of interest.

References

1. Farina, M.; Aschner, M.; Rocha, J.B. Oxidative stress in MeHg-induced neurotoxicity. *Toxicol. Appl. Pharmacol.* **2011**, *256*, 405–417. [CrossRef] [PubMed]
2. Compeau, G.C.; Bartha, R. Sulfate-reducing bacteria: Principal methylators of mercury in anoxic estuarine sediment. *Appl. Environ. Microb.* **1985**, *50*, 498–502.
3. Hintelmann, H. Organomercurials. Their formation and pathways in the environment. *Met. Ions Life Sci.* **2010**, *7*, 365–401. [PubMed]
4. Clarkson, T.W.; Magos, L.; Myers, G.J. The toxicology of mercury—Current exposures and clinical manifestations. *New Engl. J. Med.* **2003**, *349*, 1731–1737. [CrossRef] [PubMed]
5. Renieri, E.A.; Alegakis, A.K.; Kiriakakis, M.; Vinceti, M.; Ozcagli, E.; Wilks, M.F.; Tsatsakis, A.M. Cd, Pb and Hg Biomonitoring in Fish of the Mediterranean Region and Risk Estimations on Fish Consumption. *Toxics* **2014**, *2*, 417–442. [CrossRef]
6. Zareba, G.; Cernichiari, E.; Hojo, R.; Nitt, S.M.; Weiss, B.; Mumtaz, M.M.; Jones, D.E.; Clarkson, T.W. Thimerosal distribution and metabolism in neonatal mice: Comparison with methyl mercury. *J. Appl. Toxicol.* **2007**, *27*, 511–518. [CrossRef] [PubMed]
7. Costa, L.G.; Aschner, M.; Vitalone, A.; Syversen, T.; Soldin, O.P. Developmental neuropathology of environmental agents. *Annu. Rev. Pharmacol. Toxicol.* **2004**, *44*, 87–110. [CrossRef] [PubMed]
8. Hassan, S.A.; Moussa, E.A.; Abbott, L.C. The effect of methylmercury exposure on early central nervous system development in the zebrafish (Danio rerio) embryo. *J. Appl. Toxicol. JAT* **2012**, *32*, 707–713. [CrossRef] [PubMed]
9. Heimfarth, L.; Delgado, J.; Mignori, M.R.; Gelain, D.P.; Moreira, J.C.F.; Pessoa-Pureur, R. Developmental neurotoxicity of the hippocampus following in utero exposure to methylmercury: Impairment in cell signaling. *Arch. Toxicol.* **2018**, *92*, 513–527. [CrossRef] [PubMed]
10. Johansson, C.; Castoldi, A.F.; Onishchenko, N.; Manzo, L.; Vahter, M.; Ceccatelli, S. Neurobehavioural and molecular changes induced by methylmercury exposure during development. *Neurotox. Res.* **2007**, *11*, 241–260. [CrossRef] [PubMed]
11. Marsh, D.O.; Clarkson, T.W.; Myers, G.J.; Davidson, P.W.; Cox, C.; Cernichiari, E.; Tanner, M.A.; Lednar, W.; Shamlaye, C.; Choisy, O.; et al. The seychelles study of fetal methylmercury exposure and child development: Introduction. *Neurotoxicology* **1995**, *16*, 583–596. [PubMed]
12. Franco, J.L.; Teixeira, A.; Meotti, F.C.; Ribas, C.M.; Stringari, J.; Garcia Pomblum, S.C.; Moro, A.M.; Bohrer, D.; Bairros, A.V.; Dafre, A.L.; et al. Cerebellar thiol status and motor deficit after lactational exposure to methylmercury. *Environ. Res.* **2006**, *102*, 22–28. [CrossRef] [PubMed]
13. Manfroi, C.B.; Schwalm, F.D.; Cereser, V.; Abreu, F.; Oliveira, A.; Bizarro, L.; Rocha, J.B.; Frizzo, M.E.; Souza, D.O.; Farina, M. Maternal milk as methylmercury source for suckling mice: Neurotoxic effects involved with the cerebellar glutamatergic system. *Toxicol. Sci.* **2004**, *81*, 172–178. [CrossRef] [PubMed]
14. Debes, F.; Budtz-Jorgensen, E.; Weihe, P.; White, R.F.; Grandjean, P. Impact of prenatal methylmercury exposure on neurobehavioral function at age 14 years. *Neurotoxicol. Teratol.* **2006**, *28*, 536–547. [CrossRef] [PubMed]

15. Grandjean, P.; Weihe, P.; White, R.F.; Debes, F.; Araki, S.; Yokoyama, K.; Murata, K.; Sorensen, N.; Dahl, R.; Jorgensen, P.J. Cognitive deficit in 7-year-old children with prenatal exposure to methylmercury. *Neurotoxicol. Teratol.* **1997**, *19*, 417–428. [CrossRef]

16. Tatsuta, N.; Murata, K.; Iwai-Shimada, M.; Yaginuma-Sakurai, K.; Satoh, H.; Nakai, K. Psychomotor ability in children prenatally exposed to methylmercury: The 18-month follow-up of Tohoku study of child development. *Tohoku J. Exp. Med.* **2017**, *242*, 1–8. [CrossRef] [PubMed]

17. Choi, B.H. Methylmercury poisoning of the developing nervous system: I. Pattern of neuronal migration in the cerebral cortex. *Neurotoxicology* **1986**, *7*, 591–600. [PubMed]

18. Choi, B.H.; Lapham, L.W.; Amin-Zaki, L.; Saleem, T. Abnormal neuronal migration, deranged cerebral cortical organization, and diffuse white matter astrocytosis of human fetal brain: A major effect of methylmercury poisoning in utero. *J. Neuropathol. Exp. Neurol.* **1978**, *37*, 719–733. [CrossRef] [PubMed]

19. Bjorklund, O.; Kahlstrom, J.; Salmi, P.; Ogren, S.O.; Vahter, M.; Chen, J.F.; Fredholm, B.B.; Dare, E. The effects of methylmercury on motor activity are sex- and age-dependent, and modulated by genetic deletion of adenosine receptors and caffeine administration. *Toxicology* **2007**, *241*, 119–133. [CrossRef] [PubMed]

20. Carratu, M.R.; Borracci, P.; Coluccia, A.; Giustino, A.; Renna, G.; Tomasini, M.C.; Raisi, E.; Antonelli, T.; Cuomo, V.; Mazzoni, E.; et al. Acute exposure to methylmercury at two developmental windows: Focus on neurobehavioral and neurochemical effects in rat offspring. *Neuroscience* **2006**, *141*, 1619–1629. [CrossRef] [PubMed]

21. Dare, E.; Fetissov, S.; Hokfelt, T.; Hall, H.; Ogren, S.O.; Ceccatelli, S. Effects of prenatal exposure to methylmercury on dopamine-mediated locomotor activity and dopamine D2 receptor binding. *Naunyn Schmiedebergs Arch. Pharmacol.* **2003**, *367*, 500–508. [CrossRef] [PubMed]

22. Sakamoto, M.; Kakita, A.; Wakabayashi, K.; Takahashi, H.; Nakano, A.; Akagi, H. Evaluation of changes in methylmercury accumulation in the developing rat brain and its effects: A study with consecutive and moderate dose exposure throughout gestation and lactation periods. *Brain Res.* **2002**, *949*, 51–59. [CrossRef]

23. Paletz, E.M.; Craig-Schmidt, M.C.; Newland, M.C. Gestational exposure to methylmercury and n-3 fatty acids: Effects on high- and low-rate operant behavior in adulthood. *Neurotoxicol. Teratol.* **2006**, *28*, 59–73. [CrossRef] [PubMed]

24. Sfakianakis, D.G.; Renieri, E.; Kentouri, M.; Tsatsakis, A.M. Effect of heavy metals on fish larvae deformities: A review. *Environ. Res.* **2015**, *137*, 246–255. [CrossRef] [PubMed]

25. Carvalho, M.C.; Franco, J.L.; Ghizoni, H.; Kobus, K.; Nazari, E.M.; Rocha, J.B.; Nogueira, C.W.; Dafre, A.L.; Muller, Y.M.; Farina, M. Effects of 2,3-dimercapto-1-propanesulfonic acid (DMPS) on methylmercury-induced locomotor deficits and cerebellar toxicity in mice. *Toxicology* **2007**, *239*, 195–203. [CrossRef] [PubMed]

26. Farina, M.; Franco, J.L.; Ribas, C.M.; Meotti, F.C.; Missau, F.C.; Pizzolatti, M.G.; Dafre, A.L.; Santos, A.R. Protective effects of polygala paniculata extract against methylmercury-induced neurotoxicity in mice. *J. Pharm. Pharmacol.* **2005**, *57*, 1503–1508. [CrossRef] [PubMed]

27. Zimmermann, L.T.; dos Santos, D.B.; Colle, D.; dos Santos, A.A.; Hort, M.A.; Garcia, S.C.; Bressan, L.P.; Bohrer, D.; Farina, M. Methionine stimulates motor impairment and cerebellar mercury deposition in methylmercury-exposed mice. *J. Toxicol. Environ. Health A* **2014**, *77*, 46–56. [CrossRef] [PubMed]

28. Aschner, M.; Syversen, T. Methylmercury: Recent advances in the understanding of its neurotoxicity. *Ther. Drug Monit.* **2005**, *27*, 278–283. [CrossRef] [PubMed]

29. Sakamoto, M.; Nakano, A.; Kajiwara, Y.; Naruse, I.; Fujisaki, T. Effects of methyl mercury in postnatal developing rats. *Environ. Res.* **1993**, *61*, 43–50. [CrossRef] [PubMed]

30. Kaur, P.; Aschner, M.; Syversen, T. Glutathione modulation influences methyl mercury induced neurotoxicity in primary cell cultures of neurons and astrocytes. *Neurotoxicology* **2006**, *27*, 492–500. [CrossRef] [PubMed]

31. Aschner, M.; Syversen, T.; Souza, D.O.; Rocha, J.B.; Farina, M. Involvement of glutamate and reactive oxygen species in methylmercury neurotoxicity. *Braz. J. Med. Biol. Res.* **2007**, *40*, 285–291. [CrossRef] [PubMed]

32. Ceccatelli, S.; Dare, E.; Moors, M. Methylmercury-induced neurotoxicity and apoptosis. *Chem. Biol. Interact.* **2010**, *188*, 301–308. [CrossRef] [PubMed]

33. Farina, M.; Rocha, J.B.; Aschner, M. Mechanisms of methylmercury-induced neurotoxicity: Evidence from experimental studies. *Life Sci.* **2011**, *89*, 555–563. [CrossRef] [PubMed]

34. Choi, D.W. Excitotoxic cell death. *J. Neurobiol.* **1992**, *23*, 1261–1276. [CrossRef] [PubMed]

35. Lafon-Cazal, M.; Pietri, S.; Culcasi, M.; Bockaert, J. Nmda-dependent superoxide production and neurotoxicity. *Nature* **1993**, *364*, 535–537. [CrossRef] [PubMed]

36. Park, S.T.; Lim, K.T.; Chung, Y.T.; Kim, S.U. Methylmercury-induced neurotoxicity in cerebral neuron culture is blocked by antioxidants and NMDA receptor antagonists. *Neurotoxicology* **1996**, *17*, 37–45. [PubMed]

37. Farina, M.; Campos, F.; Vendrell, I.; Berenguer, J.; Barzi, M.; Pons, S.; Sunol, C. Probucol increases glutathione peroxidase-1 activity and displays long-lasting protection against methylmercury toxicity in cerebellar granule cells. *Toxicol. Sci.* **2009**, *112*, 416–426. [CrossRef] [PubMed]

38. Franco, J.L.; Posser, T.; Dunkley, P.R.; Dickson, P.W.; Mattos, J.J.; Martins, R.; Bainy, A.C.; Marques, M.R.; Dafre, A.L.; Farina, M. Methylmercury neurotoxicity is associated with inhibition of the antioxidant enzyme glutathione peroxidase. *Free Radic. Biol. Med.* **2009**, *47*, 449–457. [CrossRef] [PubMed]

39. Wagner, C.; Sudati, J.H.; Nogueira, C.W.; Rocha, J.B. In vivo and in vitro inhibition of mice thioredoxin reductase by methylmercury. *Biometals* **2010**, *23*, 1171–1177. [CrossRef] [PubMed]

40. Yin, Z.; Milatovic, D.; Aschner, J.L.; Syversen, T.; Rocha, J.B.; Souza, D.O.; Sidoryk, M.; Albrecht, J.; Aschner, M. Methylmercury induces oxidative injury, alterations in permeability and glutamine transport in cultured astrocytes. *Brain Res.* **2007**, *1131*, 1–10. [CrossRef] [PubMed]

41. Glaser, V.; Nazari, E.M.; Muller, Y.M.; Feksa, L.; Wannmacher, C.M.; Rocha, J.B.; de Bem, A.F.; Farina, M.; Latini, A. Effects of inorganic selenium administration in methylmercury-induced neurotoxicity in mouse cerebral cortex. *Int. J. Dev. Neurosci.* **2010**, *28*, 631–637. [CrossRef] [PubMed]

42. Osburn, W.O.; Kensler, T.W. Nrf2 signaling: An adaptive response pathway for protection against environmental toxic insults. *Mutat. Res.* **2008**, *659*, 31–39. [CrossRef] [PubMed]

43. Nakaso, K.; Yano, H.; Fukuhara, Y.; Takeshima, T.; Wada-Isoe, K.; Nakashima, K. Pi3k is a key molecule in the Nrf2-mediated regulation of antioxidative proteins by hemin in human neuroblastoma cells. *FEBS Lett.* **2003**, *546*, 181–184. [CrossRef]

44. Wang, L.; Chen, Y.; Sternberg, P.; Cai, J. Essential roles of the Pi3 kinase/Akt pathway in regulating Nrf2-dependent antioxidant functions in the RPE. *Investig. Ophthalmol. Vis. Sci.* **2008**, *49*, 1671–1678. [CrossRef] [PubMed]

45. Hughes, W.L. A physicochemical rationale for the biological activity of mercury and its compounds. *Ann. N. Y. Acad. Sci.* **1957**, *65*, 454–460. [CrossRef] [PubMed]

46. Naganuma, A.; Imura, N. Methylmercury binds to a low molecular weight substance in rabbit and human erythrocytes. *Toxicol. Appl. Pharmacol.* **1979**, *47*, 613–616. [CrossRef]

47. Rabenstein, D.L.; Fairhurst, M.T. Nuclear magnetic resonance studies of the solution chemistry of metal complexes. XI. The binding of methylmercury by sulfhydryl-containing amino acids and by glutathione. *J. Am. Chem. Soc.* **1975**, *97*, 2086–2092. [CrossRef] [PubMed]

48. Thomas, D.J.; Smith, J.C. Effects of coadministered low-molecular-weight thiol compounds on short-term distribution of methyl mercury in the rat. *Toxicol. Appl. Pharmacol.* **1982**, *62*, 104–110. [CrossRef]

49. Omata, S.; Sakimura, K.; Ishii, T.; Sugano, H. Chemical nature of a methylmercury complex with a low molecular weight in the liver cytosol of rats exposed to methylmercury chloride. *Biochem. Pharmacol.* **1978**, *27*, 1700–1702. [PubMed]

50. Refsvik, T.; Norseth, T. Methyl mercuric compounds in rat bile. *Acta Pharmacol. Toxicol.* **1975**, *36*, 67–78. [CrossRef]

51. Ballatori, N.; Clarkson, T.W. Biliary secretion of glutathione and of glutathione-metal complexes. *Fundam. Appl. Toxicol.* **1985**, *5*, 816–831. [CrossRef]

52. Aschner, M.; Clarkson, T.W. Uptake of methylmercury in the rat brain: Effects of amino acids. *Brain Res.* **1988**, *462*, 31–39. [CrossRef]

53. Yin, Z.; Jiang, H.; Syversen, T.; Rocha, J.B.; Farina, M.; Aschner, M. The methylmercury-L-cysteine conjugate is a substrate for the l-type large neutral amino acid transporter. *J. Neurochem.* **2008**, *107*, 1083–1090. [CrossRef] [PubMed]

54. Holben, D.H.; Smith, A.M. The diverse role of selenium within selenoproteins: A review. *J. Am. Diet. Assoc.* **1999**, *99*, 836–843. [CrossRef]

55. Papp, L.V.; Lu, J.; Holmgren, A.; Khanna, K.K. From selenium to selenoproteins: Synthesis, identity, and their role in human health. *Antioxid. Redox Signal.* **2007**, *9*, 775–806. [CrossRef] [PubMed]

56. Steinbrenner, H.; Sies, H. Protection against reactive oxygen species by selenoproteins. *Biochim. Biophys. Acta* **2009**, *1790*, 1478–1485. [CrossRef] [PubMed]

57. Lu, J.; Holmgren, A. Selenoproteins. *J. Biol. Chem.* **2009**, *284*, 723–727. [CrossRef] [PubMed]

58. Sugiura, Y.; Tamai, Y.; Tanaka, H. Selenium protection against mercury toxicity: High binding affinity of methylmercury by selenium-containing ligands in comparison with sulfur-containing ligands. *Bioinorg. Chem.* **1978**, *9*, 167–180. [CrossRef]

59. Ren, X.; Zou, L.; Zhang, X.; Branco, V.; Wang, J.; Carvalho, C.; Holmgren, A.; Lu, J. Redox signaling mediated by thioredoxin and glutathione systems in the central nervous system. *Antioxid. Redox Signal.* **2017**, *27*, 989–1010. [CrossRef] [PubMed]

60. Forman, H.J.; Zhang, H.; Rinna, A. Glutathione: Overview of its protective roles, measurement, and biosynthesis. *Mol. Aspects Med.* **2009**, *30*, 1–12. [CrossRef] [PubMed]

61. Ballatori, N.; Clarkson, T.W. Developmental changes in the biliary excretion of methylmercury and glutathione. *Science* **1982**, *216*, 61–63. [CrossRef] [PubMed]

62. Graminski, G.F.; Kubo, Y.; Armstrong, R.N. Spectroscopic and kinetic evidence for the thiolate anion of glutathione at the active site of glutathione S-transferase. *Biochemistry* **1989**, *28*, 3562–3568. [CrossRef] [PubMed]

63. Cheng, H.; Tchaikovskaya, T.; Tu, Y.S.; Chapman, J.; Qian, B.; Ching, W.M.; Tien, M.; Rowe, J.D.; Patskovsky, Y.V.; Listowsky, I.; et al. Rat glutathione s-transferase M4-4: An isoenzyme with unique structural features including a redox-reactive cysteine-115 residue that forms mixed disulphides with glutathione. *Biochem. J.* **2001**, *356*, 403–414. [CrossRef] [PubMed]

64. Khan, H.; Khan, M.F.; Jan, S.U.; Mukhtiar, M.; Ullah, N.; Anwar, N. Role of glutathione in protection against mercury induced poisoning. *Pak. J. Pharm. Sci.* **2012**, *25*, 395–400. [PubMed]

65. Patrick, L. Mercury toxicity and antioxidants: Part 1: Role of glutathione and alpha-lipoic acid in the treatment of mercury toxicity. *Altern. Med. Rev.* **2002**, *7*, 456–471. [PubMed]

66. Franco, J.L.; Braga, H.C.; Stringari, J.; Missau, F.C.; Posser, T.; Mendes, B.G.; Leal, R.B.; Santos, A.R.; Dafre, A.L.; Pizzolatti, M.G.; et al. Mercurial-induced hydrogen peroxide generation in mouse brain mitochondria: Protective effects of quercetin. *Chem. Res. Toxicol.* **2007**, *20*, 1919–1926. [CrossRef] [PubMed]

67. Ni, M.; Li, X.; Yin, Z.; Sidoryk-Wegrzynowicz, M.; Jiang, H.; Farina, M.; Rocha, J.B.; Syversen, T.; Aschner, M. Comparative study on the response of rat primary astrocytes and microglia to methylmercury toxicity. *Glia* **2011**, *59*, 810–820. [CrossRef] [PubMed]

68. Stringari, J.; Nunes, A.K.; Franco, J.L.; Bohrer, D.; Garcia, S.C.; Dafre, A.L.; Milatovic, D.; Souza, D.O.; Rocha, J.B.; Aschner, M.; et al. Prenatal methylmercury exposure hampers glutathione antioxidant system ontogenesis and causes long-lasting oxidative stress in the mouse brain. *Toxicol. Appl. Pharmacol.* **2008**, *227*, 147–154. [CrossRef] [PubMed]

69. Takeuchi, T. Pathology of minamata disease. With special reference to its pathogenesis. *Acta Pathol. Jpn.* **1982**, *32*, 73–99. [PubMed]

70. Kaur, P.; Aschner, M.; Syversen, T. Role of glutathione in determining the differential sensitivity between the cortical and cerebellar regions towards mercury-induced oxidative stress. *Toxicology* **2007**, *230*, 164–177. [CrossRef] [PubMed]

71. Wang, L.; Jiang, H.; Yin, Z.; Aschner, M.; Cai, J. Methylmercury toxicity and Nrf2-dependent detoxification in astrocytes. *Toxicol. Sci.* **2009**, *107*, 135–143. [CrossRef] [PubMed]

72. Yee, S.; Choi, B.H. Oxidative stress in neurotoxic effects of methylmercury poisoning. *Neurotoxicology* **1996**, *17*, 17–26. [PubMed]

73. Mori, K.; Yoshida, K.; Nakagawa, Y.; Hoshikawa, S.; Ozaki, H.; Ito, S.; Watanabe, C. Methylmercury inhibition of type II 5′-deiodinase activity resulting in a decrease in growth hormone production in gh3 cells. *Toxicology* **2007**, *237*, 203–209. [CrossRef] [PubMed]

74. Allen, J.W.; Shanker, G.; Aschner, M. Methylmercury inhibits the in vitro uptake of the glutathione precursor, cystine, in astrocytes, but not in neurons. *Brain Res.* **2001**, *894*, 131–140. [CrossRef]

75. Shanker, G.; Allen, J.W.; Mutkus, L.A.; Aschner, M. Methylmercury inhibits cysteine uptake in cultured primary astrocytes, but not in neurons. *Brain Res.* **2001**, *914*, 159–165. [CrossRef]

76. Aschner, M.; Yao, C.P.; Allen, J.W.; Tan, K.H. Methylmercury alters glutamate transport in astrocytes. *Neurochem. Int.* **2000**, *37*, 199–206. [CrossRef]

77. Shanker, G.; Aschner, M. Identification and characterization of uptake systems for cystine and cysteine in cultured astrocytes and neurons: Evidence for methylmercury-targeted disruption of astrocyte transport. *J. Neurosci. Res.* **2001**, *66*, 998–1002. [CrossRef] [PubMed]

78. Shanker, G.; Aschner, M. Methylmercury-induced reactive oxygen species formation in neonatal cerebral astrocytic cultures is attenuated by antioxidants. *Brain Res. Mol. Brain Res.* **2003**, *110*, 85–91. [CrossRef]

79. Kaur, P.; Aschner, M.; Syversen, T. Biochemical factors modulating cellular neurotoxicity of methylmercury. *J. Toxicol.* **2011**, *2011*, 721987. [CrossRef] [PubMed]

80. Shanker, G.; Syversen, T.; Aschner, J.L.; Aschner, M. Modulatory effect of glutathione status and antioxidants on methylmercury-induced free radical formation in primary cultures of cerebral astrocytes. *Brain Res. Mol. Brain Res.* **2005**, *137*, 11–22. [CrossRef] [PubMed]

81. Rush, T.; Liu, X.; Nowakowski, A.B.; Petering, D.H.; Lobner, D. Glutathione-mediated neuroprotection against methylmercury neurotoxicity in cortical culture is dependent on mrp1. *Neurotoxicology* **2012**, *33*, 476–481. [CrossRef] [PubMed]

82. Rayman, M.P. The importance of selenium to human health. *Lancet* **2000**, *356*, 233–241. [CrossRef]

83. Brigelius-Flohe, R.; Maiorino, M. Glutathione peroxidases. *Biochim. Biophys. Acta* **2013**, *1830*, 3289–3303. [CrossRef] [PubMed]

84. Bjornstedt, M.; Hamberg, M.; Kumar, S.; Xue, J.; Holmgren, A. Human thioredoxin reductase directly reduces lipid hydroperoxides by nadph and selenocystine strongly stimulates the reaction via catalytically generated selenols. *J. Biol. Chem.* **1995**, *270*, 11761–11764. [CrossRef] [PubMed]

85. Zhong, L.; Holmgren, A. Mammalian thioredoxin reductases as hydroperoxide reductases. *Methods Enzymol.* **2002**, *347*, 236–243. [PubMed]

86. Carvalho, C.M.; Chew, E.H.; Hashemy, S.I.; Lu, J.; Holmgren, A. Inhibition of the human thioredoxin system. A molecular mechanism of mercury toxicity. *J. Biol. Chem.* **2008**, *283*, 11913–11923. [CrossRef] [PubMed]

87. Carvalho, C.M.; Lu, J.; Zhang, X.; Arner, E.S.; Holmgren, A. Effects of selenite and chelating agents on mammalian thioredoxin reductase inhibited by mercury: Implications for treatment of mercury poisoning. *FASEB J.* **2011**, *25*, 370–381. [CrossRef] [PubMed]

88. Dalla Corte, C.L.; Wagner, C.; Sudati, J.H.; Comparsi, B.; Leite, G.O.; Busanello, A.; Soares, F.A.; Aschner, M.; Rocha, J.B. Effects of diphenyl diselenide on methylmercury toxicity in rats. *Biomed. Res. Int.* **2013**, *2013*, 983821. [CrossRef] [PubMed]

89. Ruszkiewicz, J.A.; Bowman, A.B.; Farina, M.; Rocha, J.B.T.; Aschner, M. Sex- and structure-specific differences in antioxidant responses to methylmercury during early development. *Neurotoxicology* **2016**, *56*, 118–126. [CrossRef] [PubMed]

90. Branco, V.; Canario, J.; Lu, J.; Holmgren, A.; Carvalho, C. Mercury and selenium interaction in vivo: Effects on thioredoxin reductase and glutathione peroxidase. *Free Radic. Biol. Med.* **2012**, *52*, 781–793. [CrossRef] [PubMed]

91. Penglase, S.; Hamre, K.; Ellingsen, S. Selenium prevents downregulation of antioxidant selenoprotein genes by methylmercury. *Free Radic. Biol. Med.* **2014**, *75*, 95–104. [CrossRef] [PubMed]

92. Usuki, F.; Yamashita, A.; Fujimura, M. Post-transcriptional defects of antioxidant selenoenzymes cause oxidative stress under methylmercury exposure. *J. Biol. Chem.* **2011**, *286*, 6641–6649. [CrossRef] [PubMed]

93. Fredriksson, A.; Gardlund, A.T.; Bergman, K.; Oskarsson, A.; Ohlin, B.; Danielsson, B.; Archer, T. Effects of maternal dietary supplementation with selenite on the postnatal development of rat offspring exposed to methyl mercury in utero. *Pharmacol. Toxicol.* **1993**, *72*, 377–382. [CrossRef] [PubMed]

94. Watanabe, C.; Yin, K.; Kasanuma, Y.; Satoh, H. In utero exposure to methylmercury and Se deficiency converge on the neurobehavioral outcome in mice. *Neurotoxicol. Teratol.* **1999**, *21*, 83–88. [CrossRef]

95. de Freitas, A.S.; Funck, V.R.; Rotta Mdos, S.; Bohrer, D.; Morschbacher, V.; Puntel, R.L.; Nogueira, C.W.; Farina, M.; Aschner, M.; Rocha, J.B. Diphenyl diselenide, a simple organoselenium compound, decreases methylmercury-induced cerebral, hepatic and renal oxidative stress and mercury deposition in adult mice. *Brain Res. Bull.* **2009**, *79*, 77–84. [CrossRef] [PubMed]

96. Farina, M.; Dahm, K.C.; Schwalm, F.D.; Brusque, A.M.; Frizzo, M.E.; Zeni, G.; Souza, D.O.; Rocha, J.B. Methylmercury increases glutamate release from brain synaptosomes and glutamate uptake by cortical slices from suckling rat pups: Modulatory effect of ebselen. *Toxicol. Sci.* **2003**, *73*, 135–140. [CrossRef] [PubMed]

97. Farina, M.; Frizzo, M.E.; Soares, F.A.; Schwalm, F.D.; Dietrich, M.O.; Zeni, G.; Rocha, J.B.; Souza, D.O. Ebselen protects against methylmercury-induced inhibition of glutamate uptake by cortical slices from adult mice. *Toxicol. Lett.* **2003**, *144*, 351–357. [CrossRef]

98. Roos, D.H.; Puntel, R.L.; Santos, M.M.; Souza, D.O.; Farina, M.; Nogueira, C.W.; Aschner, M.; Burger, M.E.; Barbosa, N.B.; Rocha, J.B. Guanosine and synthetic organoselenium compounds modulate methylmercury-induced oxidative stress in rat brain cortical slices: Involvement of oxidative stress and glutamatergic system. *Toxicol. In Vitro* **2009**, *23*, 302–307. [CrossRef] [PubMed]

99. Meinerz, D.F.; Branco, V.; Aschner, M.; Carvalho, C.; Rocha, J.B.T. Diphenyl diselenide protects against methylmercury-induced inhibition of thioredoxin reductase and glutathione peroxidase in human neuroblastoma cells: A comparison with ebselen. *J. Appl. Toxicol.* **2017**, *37*, 1073–1081. [CrossRef] [PubMed]

100. Cantley, L.C. The phosphoinositide 3-kinase pathway. *Science* **2002**, *296*, 1655–1657. [CrossRef] [PubMed]

101. Neri, L.M.; Borgatti, P.; Capitani, S.; Martelli, A.M. The nuclear phosphoinositide 3-kinase/Akt pathway: A new second messenger system. *Biochem. Biophys. Acta Biomembr.* **2002**, *1584*, 73–80. [CrossRef]

102. Manning, B.D.; Toker, A. AKT/PKB signaling: Navigating the network. *Cell* **2017**, *169*, 381–405. [CrossRef] [PubMed]

103. Alessi, D.R.; James, S.R.; Downes, C.P.; Holmes, A.B.; Gaffney, P.R.; Reese, C.B.; Cohen, P. Characterization of a 3-phosphoinositide-dependent protein kinase which phosphorylates and activates protein kinase balpha. *Curr. Biol.* **1997**, *7*, 261–269. [CrossRef]

104. Sarbassov, D.D.; Guertin, D.A.; Ali, S.M.; Sabatini, D.M. Phosphorylation and regulation of Akt/PKB by the rictor-mtor complex. *Science* **2005**, *307*, 1098–1101. [CrossRef] [PubMed]

105. Stambolic, V.; Suzuki, A.; de la Pompa, J.L.; Brothers, G.M.; Mirtsos, C.; Sasaki, T.; Ruland, J.; Penninger, J.M.; Siderovski, D.P.; Mak, T.W. Negative regulation of PKB/Akt-dependent cell survival by the tumor suppressor PTEN. *Cell* **1998**, *95*, 29–39. [CrossRef]

106. Das, J.; Ghosh, J.; Manna, P.; Sil, P.C. Taurine suppresses doxorubicin-triggered oxidative stress and cardiac apoptosis in rat via up-regulation of PI3-K/Akt and inhibition of p53, p38-JNK. *Biochem. Pharmacol.* **2011**, *81*, 891–909. [CrossRef] [PubMed]

107. Ohashi, H.; Takagi, H.; Oh, H.; Suzuma, K.; Suzuma, I.; Miyamoto, N.; Uemura, A.; Watanabe, D.; Murakami, T.; Sugaya, T.; et al. Phosphatidylinositol 3-kinase/Akt regulates angiotensin II-induced inhibition of apoptosis in microvascular endothelial cells by governing survivin expression and suppression of caspase-3 activity. *Circ. Res.* **2004**, *94*, 785–793. [CrossRef] [PubMed]

108. Widenmaier, S.B.; Ao, Z.; Kim, S.J.; Warnock, G.; McIntosh, C.H. Suppression of p38 MAPK and JNK via Akt-mediated inhibition of apoptosis signal-regulating kinase 1 constitutes a core component of the beta-cell pro-survival effects of glucose-dependent insulinotropic polypeptide. *J. Biol. Chem.* **2009**, *284*, 30372–30382. [CrossRef] [PubMed]

109. Hsu, A.L.; Ching, T.T.; Wang, D.S.; Song, X.; Rangnekar, V.M.; Chen, C.S. The cyclooxygenase-2 inhibitor celecoxib induces apoptosis by blocking Akt activation in human prostate cancer cells independently of Bcl-2. *J. Biol. Chem.* **2000**, *275*, 11397–11403. [CrossRef] [PubMed]

110. Huang, J.S.; Cho, C.Y.; Hong, C.C.; Yan, M.D.; Hsieh, M.C.; Lay, J.D.; Lai, G.M.; Cheng, A.L.; Chuang, S.E. Oxidative stress enhances Axl-mediated cell migration through an Akt1/Rac1-dependent mechanism. *Free Radical Bio. Med.* **2013**, *65*, 1246–1256. [CrossRef] [PubMed]

111. Luo, H.; Yang, Y.; Duan, J.; Wu, P.; Jiang, Q.; Xu, C. PTEN-regulated AKT/foxO3a/Bim signaling contributes to reactive oxygen species-mediated apoptosis in selenite-treated colorectal cancer cells. *Cell Death Dis.* **2013**, *4*, e481. [CrossRef] [PubMed]

112. Esposito, F.; Chirico, G.; Montesano Gesualdi, N.; Posadas, I.; Ammendola, R.; Russo, T.; Cirino, G.; Cimino, F. Protein kinase b activation by reactive oxygen species is independent of tyrosine kinase receptor phosphorylation and requires SRC activity. *J. Biol. Chem.* **2003**, *278*, 20828–20834. [CrossRef] [PubMed]

113. Ushio-Fukai, M.; Alexander, R.W.; Akers, M.; Yin, Q.; Fujio, Y.; Walsh, K.; Griendling, K.K. Reactive oxygen species mediate the activation of Akt/protein kinase b by angiotensin II in vascular smooth muscle cells. *J. Biol. Chem.* **1999**, *274*, 22699–22704. [CrossRef] [PubMed]

114. Gorin, Y.; Ricono, J.M.; Kim, N.H.; Bhandari, B.; Choudhury, G.G.; Abboud, H.E. Nox4 mediates angiotensin ii-induced activation of Akt/protein kinase b in mesangial cells. *Am. J. Physiol. Renal* **2003**, *285*, F219–F229. [CrossRef] [PubMed]

115. Sonoda, Y.; Watanabe, S.; Matsumoto, Y.; Aizu-Yokota, E.; Kasahara, T. Fak is the upstream signal protein of the phosphatidylinositol 3-kinase-Akt survival pathway in hydrogen peroxide-induced apoptosis of a human glioblastoma cell line. *J. Biol. Chem.* **1999**, *274*, 10566–10570. [CrossRef] [PubMed]

116. Fujimura, M.; Usuki, F. Methylmercury causes neuronal cell death through the suppression of the Trka pathway: In vitro and in vivo effects of Trka pathway activators. *Toxicol. Appl. Pharm.* **2015**, *282*, 259–266. [CrossRef] [PubMed]

117. Pierozan, P.; Biasibetti, H.; Schmitz, F.; Avila, H.; Fernandes, C.G.; Pessoa-Pureur, R.; Wyse, A.T.S. Neurotoxicity of methylmercury in isolated astrocytes and neurons: The cytoskeleton as a main target. *Mol. Neurobiol.* **2017**, *54*, 5752–5767. [CrossRef] [PubMed]

118. Chen, Y.W.; Huang, C.F.; Tsai, K.S.; Yang, R.S.; Yen, C.C.; Yang, C.Y.; Lin-Shiau, S.Y.; Liu, S.H. The role of phosphoinositide 3-kinase/Akt signaling in low-dose mercury-induced mouse pancreatic beta-cell dysfunction in vitro and in vivo. *Diabetes* **2006**, *55*, 1614–1624. [CrossRef] [PubMed]

119. Unoki, T.; Abiko, Y.; Toyama, T.; Uehara, T.; Tsuboi, K.; Nishida, M.; Kaji, T.; Kumagai, Y. Methylmercury, an environmental electrophile capable of activation and disruption of the Akt/CREB/Bcl-2 signal transduction pathway in SH-SY5Y cells. *Sci. Rep.* **2016**, *6*, 28944. [CrossRef] [PubMed]

120. Culbreth, M.; Zhang, Z.; Aschner, M. Methylmercury augments Nrf2 activity by downregulation of the Src family kinase Fyn. *Neurotoxicology* **2017**, *62*, 200–206. [CrossRef] [PubMed]

121. Silva-Islas, C.A.; Maldonado, P.D. Canonical and non-canonical mechanisms of Nrf2 activation. *Pharmacol. Res.* **2018**, *134*, 92–99. [CrossRef] [PubMed]

122. Kobayashi, M.; Yamamoto, M. Molecular mechanisms activating the Nrf2-Keap1 pathway of antioxidant gene regulation. *Antioxid. Redox Signal.* **2005**, *7*, 385–394. [CrossRef] [PubMed]

123. Canning, P.; Sorrell, F.J.; Bullock, A.N. Structural basis of keap1 interactions with Nrf2. *Free Radic. Biol. Med.* **2015**, *88*, 101–107. [CrossRef] [PubMed]

124. Kaspar, J.W.; Niture, S.K.; Jaiswal, A.K. Nrf2:Inrf2 (Keap1) signaling in oxidative stress. *Free Radic. Biol. Med.* **2009**, *47*, 1304–1309. [CrossRef] [PubMed]

125. Huang, H.C.; Nguyen, T.; Pickett, C.B. Phosphorylation of Nrf2 at Ser-40 by protein kinase c regulates antioxidant response element-mediated transcription. *J. Biol. Chem.* **2002**, *277*, 42769–42774. [CrossRef] [PubMed]

126. Joo, M.S.; Kim, W.D.; Lee, K.Y.; Kim, J.H.; Koo, J.H.; Kim, S.G. Ampk facilitates nuclear accumulation of Nrf2 by phosphorylating at serine 550. *Mol. Cell. Biol.* **2016**, *36*, 1931–1942. [CrossRef] [PubMed]

127. Calkins, M.J.; Johnson, D.A.; Townsend, J.A.; Vargas, M.R.; Dowell, J.A.; Williamson, T.P.; Kraft, A.D.; Lee, J.M.; Li, J.; Johnson, J.A. The Nrf2/are pathway as a potential therapeutic target in neurodegenerative disease. *Antioxid. Redox Signal.* **2009**, *11*, 497–508. [CrossRef] [PubMed]

128. de Oliveira, M.R.; Ferreira, G.C.; Schuck, P.F.; Dal Bosco, S.M. Role for the Pi3K/Akt/Nrf2 signaling pathway in the protective effects of carnosic acid against methylglyoxal-induced neurotoxicity in SH-SY5Y neuroblastoma cells. *Chem. Biol. Interact.* **2015**, *242*, 396–406. [CrossRef] [PubMed]

129. Wang, L.; Zhang, S.; Cheng, H.; Lv, H.; Cheng, G.; Ci, X. Nrf2-mediated liver protection by esculentoside a against acetaminophen toxicity through the AMPK/Akt/GSK3β pathway. *Free Radic. Biol. Med.* **2016**, *101*, 401–412. [CrossRef] [PubMed]

130. Niture, S.K.; Khatri, R.; Jaiswal, A.K. Regulation of Nrf2-an update. *Free Radic Biol Med* **2014**, *66*, 36–44. [CrossRef] [PubMed]

131. Cuadrado, A. Structural and functional characterization of Nrf2 degradation by glycogen synthase kinase 3/β-TrCP. *Free Radic. Biol. Med.* **2015**, *88*, 147–157. [CrossRef] [PubMed]

132. Rada, P.; Rojo, A.I.; Chowdhry, S.; McMahon, M.; Hayes, J.D.; Cuadrado, A. SCF/{beta}-TrCP promotes glycogen synthase kinase 3-dependent degradation of the Nrf2 transcription factor in a keap1-independent manner. *Mol. Cell. Biol.* **2011**, *31*, 1121–1133. [CrossRef] [PubMed]

133. Chowdhry, S.; Zhang, Y.; McMahon, M.; Sutherland, C.; Cuadrado, A.; Hayes, J.D. Nrf2 is controlled by two distinct β-TrCP recognition motifs in its Neh6 domain, one of which can be modulated by GSK-3 activity. *Oncogene* **2013**, *32*, 3765–3781. [CrossRef] [PubMed]

134. Rada, P.; Rojo, A.I.; Evrard-Todeschi, N.; Innamorato, N.G.; Cotte, A.; Jaworski, T.; Tobon-Velasco, J.C.; Devijver, H.; Garcia-Mayoral, M.F.; Van Leuven, F.; et al. Structural and functional characterization of Nrf2 degradation by the glycogen synthase kinase 3/beta-trcp axis. *Mol. Cell. Biol.* **2012**, *32*, 3486–3499. [CrossRef] [PubMed]

135. Jain, A.K.; Jaiswal, A.K. Gsk-3beta acts upstream of Fyn kinase in regulation of nuclear export and degradation of NF-E2 related factor 2. *J. Biol. Chem.* **2007**, *282*, 16502–16510. [CrossRef] [PubMed]

136. Toyama, T.; Sumi, D.; Shinkai, Y.; Yasutake, A.; Taguchi, K.; Tong, K.I.; Yamamoto, M.; Kumagai, Y. Cytoprotective role of Nrf2/Keap1 system in methylmercury toxicity. *Biochem. Biophys. Res. Commun.* **2007**, *363*, 645–650. [CrossRef] [PubMed]

137. Feng, S.; Xu, Z.; Wang, F.; Yang, T.; Liu, W.; Deng, Y.; Xu, B. Sulforaphane prevents methylmercury-induced oxidative damage and excitotoxicity through activation of the Nrf2-ARE pathway. *Mol. Neurobiol.* **2017**, *54*, 375–391. [CrossRef] [PubMed]

138. Toyama, T.; Shinkai, Y.; Yasutake, A.; Uchida, K.; Yamamoto, M.; Kumagai, Y. Isothiocyanates reduce mercury accumulation via an Nrf2-dependent mechanism during exposure of mice to methylmercury. *Environ. Health Perspect.* **2011**, *119*, 1117–1122. [CrossRef] [PubMed]

139. Toyama, T.; Shinkai, Y.; Kaji, T.; Kumagai, Y. Convenient method to assess chemical modification of protein thiols by electrophilic metals. *J. Toxicol. Sci.* **2013**, *38*, 477–484. [CrossRef] [PubMed]

140. Kumagai, Y.; Kanda, H.; Shinkai, Y.; Toyama, T. The role of the Keap1/Nrf2 pathway in the cellular response to methylmercury. *Oxid. Med. Cell. Longev.* **2013**, *2013*, 848279. [CrossRef] [PubMed]

141. Yoshida, E.; Abiko, Y.; Kumagai, Y. Glutathione adduct of methylmercury activates the Keap1-Nrf2 pathway in SH-SY5Y cells. *Chem. Res. Toxicol.* **2014**, *27*, 1780–1786. [CrossRef] [PubMed]

142. Levonen, A.L.; Landar, A.; Ramachandran, A.; Ceaser, E.K.; Dickinson, D.A.; Zanoni, G.; Morrow, J.D.; Darley-Usmar, V.M. Cellular mechanisms of redox cell signalling: Role of cysteine modification in controlling antioxidant defences in response to electrophilic lipid oxidation products. *Biochem. J.* **2004**, *378*, 373–382. [CrossRef] [PubMed]

143. Taguchi, K.; Motohashi, H.; Yamamoto, M. Molecular mechanisms of the Keap1-Nrf2 pathway in stress response and cancer evolution. *Genes Cells* **2011**, *16*, 123–140. [CrossRef] [PubMed]

144. Gao, P.; Li, L.; Ji, L.; Wei, Y.; Li, H.; Shang, G.; Zhao, Z.; Chen, Q.; Jiang, T.; Zhang, N. Nrf2 ameliorates diabetic nephropathy progression by transcriptional repression of TGFβ1 through interactions with c-Jun and SP1. *Biochim. Biophys. Acta* **2014**, *1839*, 1110–1120. [CrossRef] [PubMed]

toxics

MDPI

Review

Assessment of Cardiac Autonomic Function in Relation to Methylmercury Neurotoxicity

Kanae Karita [1], Toyoto Iwata [2], Eri Maeda [2], Mineshi Sakamoto [3] and Katsuyuki Murata [2,*]

[1] Department of Hygiene and Public Health, Kyorin University School of Medicine, Mitaka, Tokyo 181-8611, Japan; kanae@ks.kyorin-u.ac.jp

[2] Department of Environmental Health Sciences, Akita University Graduate School of Medicine, Akita, Akita 010-8543, Japan; iwata@med.akita-u.ac.jp (T.I.); erimaeda@med.akita-u.ac.jp (E.M.)

[3] National Institute for Minamata Disease, Minamata, Kumamoto 867-0008, Japan; sakamoto@nimd.go.jp

* Correspondence: winestem@med.akita-u.ac.jp; Tel.: +81-18-884-6082

Received: 2 June 2018; Accepted: 18 July 2018; Published: 20 July 2018

Abstract: After the European Food Safety Authority reviewed reports of methylmercury and heart rate variability (HRV) in 2012, the panel concluded that, although some studies of cardiac autonomy suggested an autonomic effect of methylmercury, the results were inconsistent among studies and the implications for health were unclear. In this study, we reconsider this association by adding a perspective on the physiological context. Cardiovascular rhythmicity is usually studied within different frequency domains of HRV. Three spectral components are usually detected; in humans these are centered at <0.04 Hz, 0.15 Hz (LF), and 0.3 Hz (HF). LF and HF (sympathetic and parasympathetic activities, respectively) are evaluated in terms of frequency and power. By searching PubMed, we identified 13 studies examining the effect of methylmercury exposure on HRV in human populations in the Faroe Islands, the Seychelles and other countries. Considering both reduced HRV and sympathodominant state (i.e., lower HF, higher LF, or higher LF/HF ratio) as autonomic abnormality, eight of them showed the significant association with methylmercury exposure. Five studies failed to demonstrate any significant association. In conclusion, these data suggest that increased methylmercury exposure was consistently associated with autonomic abnormality, though the influence of methylmercury on HRV (e.g., LF) might differ for prenatal and postnatal exposures. The results with HRV should be included in the risk characterization of methylmercury. The HRV parameters calculated by frequency domain analysis appear to be more sensitive to methylmercury exposure than those by time domain analysis.

Keywords: heart rate variability; methylmercury neurotoxicity; review; sympathodominant state

1. Introduction

The measurement of heart rate variability (HRV; or, the coefficient of variation of the R-R intervals, CV_{RR}) using frequency domain analysis is an effective approach for the objective assessment of the autonomic nervous function [1–3]. In a seminal study, Wheeler and Watkins observed a striking reduction or absence of beat-to-beat variation during both quiet and deep breathing in diabetic patients with autonomic neuropathy [4]. In 2012, the European Food Safety Authority (EFSA) reviewed several reports examining the influence of methylmercury on HRV and concluded that, although some studies of cardiac autonomy suggested an autonomic nervous effect of methylmercury, the results were inconsistent across studies and the implications for health were unclear [5]. Gribble et al. [6] reached a similar conclusion, finding that a major limitation of studies examining the influence of methylmercury on HRV was a lack of standardized methods for performing and reporting HRV measurements. Concerning the implications of HRV findings for health, however, the existing data may allow other interpretations besides those drawn by the EFSA panel. This study was intended to reconsider the involvement of methylmercury in HRV by adding a perspective on the physiological context.

2. Materials and Methods

2.1. Data Sources and Extraction

We searched for published papers using a set of keywords ("mercury," "heart rate variability," and "humans") in PubMed, US National Library of Medicine, and identified 33 citations as of 31 January 2018. We excluded review papers ($n = 6$), case series ($n = 3$), studies addressing occupational mercury exposure ($n = 2$) or studies in which data for either HRV ($n = 3$) or mercury biomarkers ($n = 6$) were not described. Finally, included were 13 studies of methylmercury exposure and HRV: ten studies [7–16] that were included in a previous systematic review by Gribble et al. [6] and three new studies [17–19]. In all of the identified studies, the statistical significance was set at $p < 0.05$.

2.2. Physiological Background

The autonomic nervous system innervates every organ in the body, and its neural organization in the brain, spinal cord, and periphery is as complex as the somatic nervous system [1]. A vasomotor center, located in the medulla oblongata, and vagal cardioinhibitory neurons, located primarily within the ventrolateral subdivision of the nucleus ambiguus, are thought to be especially important in the regulation of the cardiovascular system. The autonomic nervous system plays a role in triggering or sustaining malignant ventricular arrhythmias. Higher sympathetic activity, unopposed by vagal activity, promotes arrhythmia through a variety of mechanisms such as reducing the ventricular refractory period and the ventricular fibrillation threshold, promoting triggered activity afterpotentials, and enhancing automaticity [20]. By contrast, vagal stimulation opposes these changes and reduces the effects of sympathetic stimulation by prolonging refractoriness, elevating the ventricular fibrillation threshold, and reducing automaticity. For this reason, HRV testing is important for assessing cardiac autonomic function in clinical applications because of the availability of low-cost and non-invasive methods.

The procedure for data sampling and spectral analysis of successive R-R intervals on electrocardiograph (ECG) is illustrated in Figure 1. The sinus rhythm shows fluctuation around the mean R-R interval (or heart rate) due to continuous changes in the sympathovagal balance [21]. Rhythmicity is usually studied within different frequency domains. Three major spectral components, calculated by frequency domain analysis using a fast Fourier transform (FFT) or autoregressive model, are usually detected. In humans, these are centered at a very low frequency (VLF—below 0.04 Hz), a low frequency (LF—around 0.15 Hz) and a high frequency (HF—around 0.3 Hz) [1]. Spectral analysis involves subjecting a time series of R-R intervals to a mathematical transformation which separates those R-R intervals into individual harmonics which are identifiable through their discrete frequencies. The LF and HF are evaluated in terms of frequency and amplitude; the latter commonly assessed by its area (i.e., power spectral density or colloquially "power"). The VLF has been equated with a thermoregulatory or vasomotor influence, the LF with baroreflex control and arterial pressure variations, and the HF with respiration [1,21]. Therefore, the LF and HF are thought to be mediated by sympathetic and parasympathetic pathways, respectively [1,2]. The CV_{RR} was defined as the ratio of the standard deviation (SD) of the R-R intervals to the average value (RR_{mean}). Likewise, the CV_{LF} and CV_{HF} were defined as the ratios of the square roots of each component power spectral density (i.e., LF and HF) to the RR_{mean}, which can be compared between different populations because they are adjusted for the mean R-R interval of each subject [3]. In addition, the LF/HF ratio and %LF (= $LF/(LF + HF) \times 100$, %) are used as HRV parameters for frequency domain metrics.

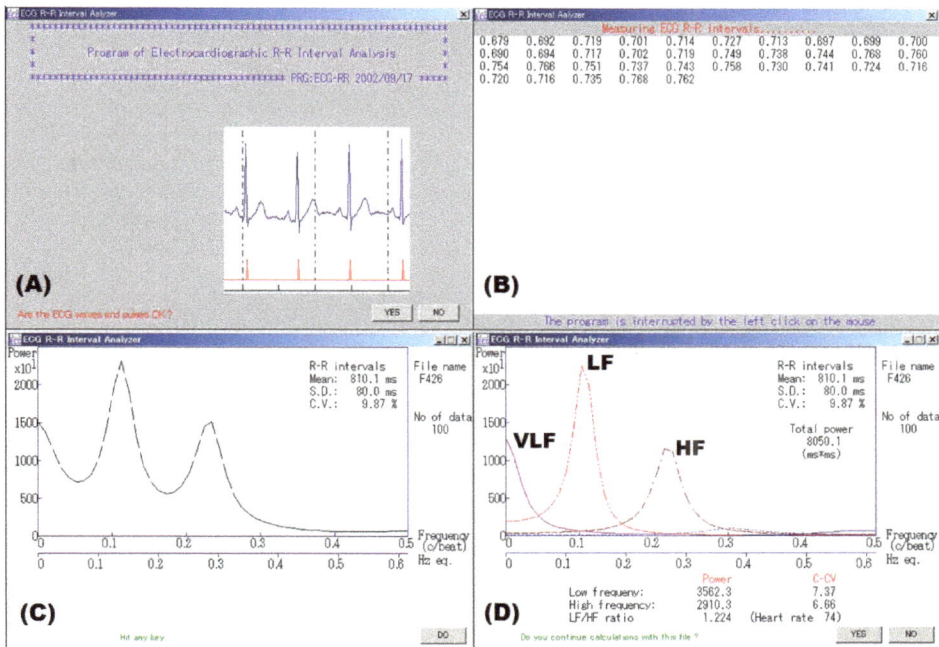

Figure 1. An example of spectral analysis of ECG R-R intervals. (**A**) An examiner confirmed waveforms on the electrocardiogram. (**B**) Electrocardiographic 300 R-R intervals or 5-min R-R intervals were measured in real time. (**C**) Spectral analysis using autoregressive model was made for consecutive 100 R-R intervals with the minimal standard deviation (SD) that were automatically extracted from the obtained data, and (**D**) component analysis was made for the data. Finally, a coefficient of variation (C.V. or CV_{RR}) was calculated from the mean and SD of the same data. VLF, LF, and HF represent very low frequency, low frequency and high frequency bands, respectively.

2.3. Interpretation of HRV Parameters

Autonomic abnormality is thought to represent a sympathodominant state of the sympathovagal balance during the initial stages and depressed HRV at the severe stage. The latter manifests as a reduction in total (e.g., CV_{RR}) and in specific power (i.e., HF, LF, CV_{HF}, and CV_{LF}) of spectral components and is observed in patients after acute myocardial infarction [2], and those with autonomic neuropathy due to diabetes mellitus [1] or alcoholism [22]. The sympathodominant state has three patterns compared with healthy controls: (i) lower HF (with no significant difference in LF); (ii) higher LF (with no significant difference in HF); and (iii) higher LF/HF ratio (or %LF). Previous studies on the effects of occupational and environmental factors on HRV parameters have primarily demonstrated a lower HF pattern (due to lead, styrene, mixed solvents such as *n*-hexane and toluene; local vibration in chain-saw workers [23–30]; exposure to sarin [31]; and long commuting times of 90 min or more [32]). In addition, Pagani et al. [33] observed a higher LF at rest in patients diagnosed with chronic fatigue syndrome in comparison with healthy control subjects (73 ± 11 and 51 ± 10 normalized units, respectively, $p < 0.05$), but responsiveness to mental stimuli (mental arithmetic) was reduced in the patients compared with the controls. Thus, the above interpretation may be applicable for HRV assessment in a static state, but not in an active mode, because the latter cannot preserve stationarity of autonomic modulations.

3. Results

3.1. Relations of Methylmercury to HRV

Table 1 shows the characteristics of epidemiological studies examining the effects of methylmercury exposure on HRV in chronological order, together with the range of exposure levels (i.e., minimum and maximum) for the studies that provided this information. In one case-control study, patients officially certified as having fetal-type Minamata disease (methylmercury poisoning due to *in utero* exposure) showed significantly reduced HF compared to age- and sex-matched healthy controls, but no data on prenatal or postnatal methylmercury exposure were included [7].

Table 1. Human studies addressing the effects of methylmercury exposure on heart rate variability.

Authors (Year) [Ref #]	Place	Subjects	Prenatal Exposure (Total Mercury Levels) *	Postnatal Exposure (Total Mercury Levels)
Oka et al. (2003) [7]	Minamata, Japan	9 FMD patients and 13 controls		
Grandjean et al. (2004) [8]	Faroe Islands, Denmark	857 children aged 7 years	GM 22.6 µg/L, IQR 13.2~40.8 µg/L in cord blood; GM 4.22 µg/g, IQR 2.55~7.68 µg/g in maternal hair	GM 2.99 µg/g, IQR 1.69~6.20 µg/g in hair
		857 children aged 14 years		GM 0.96 µg/g, IQR 0.45~2.29 µg/g in hair
Murata et al. (2006) [9]	Japan	136 children	Med 0.089 µg/g, range 0.017~0.367 µg/g in cord tissue	Med 1.66 µg/g, range 0.43~6.32 µg/g in hair
Valera et al. (2008) [10]	Nunavik, Canada	205 Inuit adults		GM 19.6 µg/L, range 0.5~152 µg/L in blood
Choi et al. (2009) [11]	Faroe Islands, Denmark	42 whaling men		GM 7.31 µg/g, IQR 4.52~13.4 µg/g in hair; GM 29.5 µg/L, IQR 18.7~46.1 µg/L in blood
Yaginuma-Sakurai et al. (2010) [12]	Sendai, Japan	Intervention group (IG): 27 adults Control group (CG): 27 adults		IG: 2.30 ± 1.08 µg/g (Mean ± SD, 0th week), 8.76 ± 2.01 µg/g (15th week); CG: 2.27 ± 1.2 µg/g (0th week), 2.14 ± 1.03 µg/g (15th week) in hair
Lim et al. (2010) [13]	South Korea	1589 adults		GM 0.83 µg/g, IQR 0.56~1.28 µg/g in hair
Valera et al. (2011) [14]	Quebec, Canada	724 Cree adults		Med 5.7 µg/L, IQR 1.2~8.8 µg/L in blood
Valera et al. (2011) [15]	French Polynesia	101 teenagers		Med 8.5 µg/L, IQR 6.3~11.0 µg/L in blood
		180 adults		Med 13.5 µg/L, IQR 8.5~22.0 µg/L in blood
Valera et al. (2012) [16]	Nunavik, Canada	226 Inuit children	Med 16.3 µg/L, IQR 9.0~28.0 µg/L in cord blood	Med 2.9 µg/L, IQR 1.5~5.6 µg/L in blood
Periard et al. (2015) [17]	Seychelles	95 adolescents	Mean 6.7 µg/g, range 0.7~21.3 µg/g in maternal hair	Mean 9.5 µg/g, range 2.0~28.1 µg/g in hair
Gump et al. (2017) [18]	Syracuse, NY, USA	203 children		Mean 0.4 µg/L, range 0.01~11.65 µg/L in blood
Miller et al. (2017) [19]	Long Island, NY, USA	94 fish consumers		8.4 ± 8.6 (Mean ± SD) µg/L in blood

* Methylmercury levels were measured only in cord tissue [26]. *Abbreviations:* FMD, fetal-type Minamata disease; GM, geometric mean value; IQR, interquartile range (25th and 75th percentiles); Med, median value.

In the Faroese birth cohort study, cord-blood mercury was associated with decreased LF and HF in children aged 14 years [8], and there were significant associations between increased mercury levels in cord blood and hair at 7 years and decreased LF and CV_{LF} in the children aged 7 years. In a retrospective cohort study using dry cord tissue, methylmercury levels in cord tissue were associated with increased LF/HF ratio and decreased HF in Japanese children at 7 years of age [9]. The median mercury level in this population was estimated to be 2.24 (range, 0.43–9.26) µg/g in maternal hair at parturition according to the equation of Akagi et al. [34]. In 11-year-old Inuit children, blood mercury was associated with decreased CV_{RR} (adjusted $\beta = -0.06$, $p = 0.01$) and LF (adjusted $\beta = -0.24$, $p = 0.02$), though neither cord-blood mercury nor hair mercury at 11 years was significantly associated with any HRV parameter [16]. The Seychelles child development study found no significant associations between prenatal or postnatal exposure to methylmercury and HRV parameters [17].

In a cross-sectional study, Lim et al. [13] reported that hair mercury was negatively related to HF in Korean adults ($p < 0.05$). In Cree adults, blood and hair mercury levels were positively related to LF/HF ratio, LF, and HF ($p < 0.01$) [14]. Similarly, in French Polynesians aged 12–17 years, significant differences were observed in LF/HF ratio, LF, and HF between the second (7.9–10.0 µg/L) and third (11.0–26.0 µg/L) tertiles of blood mercury concentration [15]. Among Faroese whaling men, blood mercury level was associated with increased CV_{RR}, CV_{HR}, and CV_{LF}, but latent mercury level, estimated from mercury levels in blood, toe nail, and hair (7 years ago) using a structural equation model, was not significantly associated with any HRV parameter [11]. In Inuit adults, blood mercury was significantly correlated with CV_{RR} and LF ($r = -0.18$ and $r = -0.18$, respectively), but these significant associations disappeared after adjusting for potential confounders [10]. In avid fish consumers, either blood total mercury or serum docosahexaenoic acid (DHA) and eicosapentaenoic acid (EPA) level was not significantly associated with any HRV parameter in a multiple regression analysis [19]. Gump et al. [18] measured LF, HF, and LF/HF ratio at rest and during stress in children aged 9–11 years to assess parasympathetic responses to acute stress, but neither blood mercury nor lead was significantly related to baseline HRV parameters (HRV data not shown).

An intervention study reported that methylmercury exposure of 3.4 µg/kg body weight/week for 14 weeks via fish consumption induced a temporary sympathodominant state ($p = 0.014$ for higher CV_{LF}; $p = 0.076$ for elevated LF/HF ratio) [12]. Also, age-, sex-, and body mass index-adjusted CV_{LF} was positively related to hair mercury at the 15th week ($p < 0.001$), though such significant relations were not observed at baseline or at the 29th week of follow-up. Table 2 presents a summary of studies examining the association between mercury levels and HRV parameters. A detailed discussion of these results follows in Section 4.1.

Table 2. Summary of associations between mercury levels and heart rate variability (HRV) parameters.

Authors (Year) [Ref #]	Mean Age at the Time of Examination	Exposure Period	HRV Parameters			
			CV_{RR}	HF-Related Parameters	LF-Related Parameters	LF/HF Ratio
Oka et al. (2003) [7]	Patients 44.3 years, controls 42.9 years	prenatal	c(±)	c(-)	c(±)	
Grandjean et al. (2004) [8]	7 years	prenatal	r(±)	r(±)	r(-)	r(±)
		postnatal	r(±)	r(±)	r(-)	r(±)
	14 years	prenatal	r(-)	r(-)	r(-)	r(±)
		postnatal	r(±)	r(±)	r(±)	r(±)
Murata et al. (2006) [9]	6.9 years	prenatal		r(-)	r(±)	r(+)
		postnatal		r(±)	r(+)	r(±)
Valera et al. (2008) [10]	52.1 years	postnatal	r(±)	r(±)	r(±)	r(±)
Choi et al. (2009) [11]	58.9 years	postnatal	r(±)	r(±)	r(±)	
Yaginuma-Sakurai et al. (2010) [12]	Intervention 25.2 years; control 23.7 years	postnatal	c(±)	c(±)	c(+)	c(±)
			r(±)	r(±)	r(+)	r(±)
Lim et al. (2010) [13]	33 years	postnatal		r(-)	r(±)	
Valera et al. (2011) [14]	35 years	postnatal		r(+)	r(+)	r(+)
Valera et al. (2011) [15]	14.2 years	postnatal		c(-)	c(+)	c(+)
	48.6 years	postnatal		c(±)	c(±)	c(±)
Valera et al. (2012) [16]	11.3 years	prenatal	r(±)	r(±)	r(±)	r(±)
		postnatal	r(-)	r(±)	r(-)	r(±)
Periard et al. (2015) [17]	19.5 years	prenatal		r(±)		r(±)
		postnatal		r(±)		r(±)
Gump et al. (2017) [18]	10.6 years	postnatal		r(±)	r(±)	r(±)
Miller et al. (2017) [19]	48.9 years	postnatal		r(±)	r(±)	r(±)

Notes: c(-), significantly low in comparison; c(+), significantly high in comparison; c(±), not significant in comparison; r(-), significantly negative relation; r(+), significantly positive relation; r(±), no significant relation. Gray areas show a sympathodominant state or autonomic hypofunction.

3.2. Measurement of HRV

HRV has been analyzed using a Holter monitoring system [10,14–16,19], an ECG analyzer with analog-to-digital converter [7–9,12], or other ECG measuring instruments [11,13,17,18]. ECG signals for R-R intervals were digitalized at 128 Hz [10,14–16], 200 Hz [17], 250 Hz [7], 500 Hz [18], and 1000 Hz [8,9,12]. Three reports did not mention the sampling frequency [11,13,19]. Most of the study subjects were examined in a supine position [7–9,11,12,17] after subjects lay quietly for two min or longer, but the subjects examined by Lim et al. [13] were sitting in a quiet and dark room. The remaining reports did not provide any detailed information [10,14–16,18,19].

To examine the effect of sampling frequency on LF and HF, 128 consecutive R-R intervals with the minimal SD in 61 male students aged 18–26 years [35] were reanalyzed using FFT spectral analysis, as shown in Figure 2. The original R-R intervals were measured at a sampling frequency of 1000 Hz after each subject rested in a supine position for 10 min, and data for lower sampling frequencies (500 Hz, 250 Hz, 200 Hz, 125 Hz, and 100 Hz) were generated from the original data taking into account random error. Table 3 presents the HRV parameters calculated by the spectral analysis. All the HRV parameters were significantly different among six frequency-band groups. In particular, most of significant differences were observed between sampling frequency bands of less than 200 Hz and 200–1000 Hz.

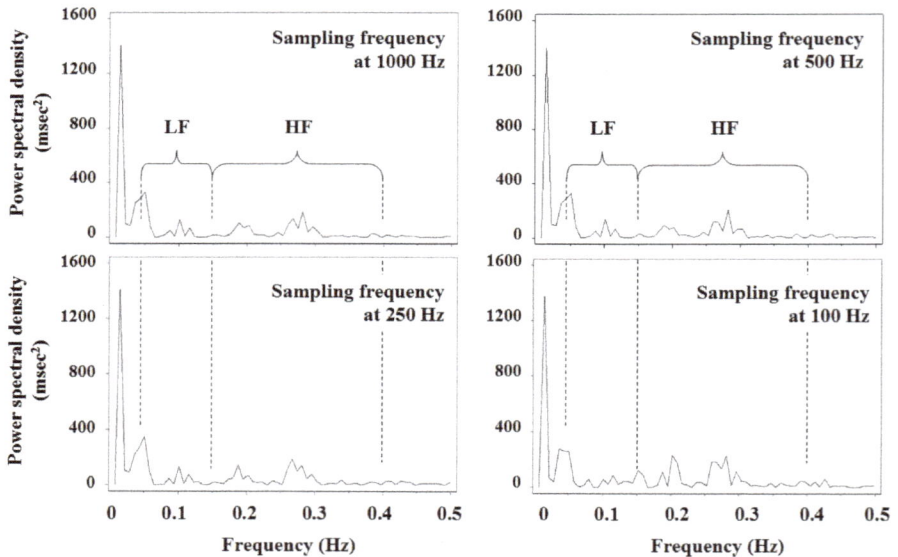

Figure 2. Results of spectral analysis of 128 consecutive R-R intervals using a fast Fourier transform in a male student. The original R-R intervals were measured at the sampling frequency of 1000 Hz after a subject rested in the supine position for 10 min, and data for lower sampling frequencies (500 Hz, 250 Hz and 100 Hz) were generated from the original data taking into account random error. LF and HF represent low frequency and high frequency bands, respectively.

Table 3. Mean +/- SD values of HRV parameters measured at different sampling frequencies in 61 male students of reference [35].

Parameters	Sampling Frequency					
	1000 Hz	500 Hz	250 Hz	200 Hz	125 Hz	100 Hz
RR_{mean} (msec)	965.1 ± 162.9	965.1 ± 162.9	965.1 ± 162.9	965.1 ± 162.9	965.2 ± 162.9 *	965.2 ± 162.9
RR_{SD} (msec)	51.74 ± 21.35	51.75 ± 21.35	51.75 ± 21.34	51.75 ± 21.33	51.82 ± 21.29 *	51.80 ± 21.27 *
CV_{RR} (%)	5.321 ± 1.902	5.322 ± 1.903	5.323 ± 1.903	5.323 ± 1.903	5.330 ± 1.896 *	5.328 ± 1.896 *
$\log_{10}$ [LF (msec^2)]	4.862 ± 0.480	4.862 ± 0.480	4.863 ± 0.478	4.863 ± 0.478	4.864 ± 0.478	4.858 ± 0.486 *
$\log_{10}$ [HF (msec^2)]	4.931 ± 0.537	4.931 ± 0.537	4.932 ± 0.534	4.932 ± 0.533	4.936 ± 0.525	4.938 ± 0.521 *
%LF (%)	46.91 ± 20.01	46.89 ± 20.00	46.88 ± 19.95	46.88 ± 19.96	46.68 ± 20.06	46.33 ± 20.14 *
$\log_{10}$ [LF/HF ratio]	−0.069 ± 0.416	−0.069 ± 0.416	−0.069 ± 0.415	−0.069 ± 0.415	−0.073 ± 0.416	−0.081 ± 0.420 *

Notes: LF (low frequency) and HF (high frequency) powers were calculated by spectral analysis shown in Figure 2; * shows $p < 0.05$ of significance levels obtained by two-way analysis of variance (F test) adding data of a lower sampling-frequency band stepwise.

4. Discussion

4.1. Assessment of Methylmercury Neurotoxicity

With regard to the interpretation of HRV at rest, depressed HRV (e.g., CV_{RR} of less than 2%) takes precedence of a sympathodominant state because it indicates cardiac autonomic hypofunction while other readouts reflect a result of disrupted sympathovagal balance [21,36]. Empirically, reduced HF precedes elevated LF and LF/HF ratio, as mentioned in Section 2.3. Furthermore, elevated LF may precede high LF/HF ratio because the ratio is a relative measure of sympathetic and parasympathetic nerve activities and is not always suggestive of autonomic dysfunction [37]. Of course, data that did not achieve statistical significance (c(±) or r(±)) are no longer discussed because any marginal association was likely attributable to chance. In light of these criteria, some commonalities can be observed across the 13 studies examined here, though Gribble et al. [6] judged the above evidence was too limited to draw causal inferences. Namely, increased mercury levels were associated with autonomic abnormality as shown in Table 2 (especially, gray areas). In addition, some fetal-type and child Minamata disease patients showed vegetative symptoms including dizziness, orthostatic syncope, palpitation, breathlessness, and nausea [38], along with hypersalivation and ileus [7].

Of the eight studies showing significant associations in Table 2, three demonstrated a potential causal link between prenatal exposure level of methylmercury and autonomic hypofunction [7–9]; this finding is similar to some results for cognitive deficits [39,40] and mental retardations [41]. Likewise, five studies suggested a significant relationship existed between postnatal exposure levels and HRV parameters analyzed using the frequency domain method. It is relatively straightforward to infer causal relations from cohort studies [8,9] and an intervention study [12]. Contrariwise, it is difficult to discriminate the effects of postnatal exposure to methylmercury from those of prenatal exposure based on cross-sectional studies with no information about prenatal exposure levels [13–15], inasmuch as there is evidence that mercury levels at 7 years of age reflect the prenatal exposure levels to some extent [42]. In cases of this nature, it would be necessary to develop a plan estimating prenatal exposure levels: For example, current mercury levels in hair of mothers, who had not changed their dietary habits, might be used as a proxy for mercury exposure during pregnancy [43,44]. This method cannot be applicable to subjects aged more than 7 years, as mentioned above. Insignificant findings may have been attributable to extremely low exposure levels [18,19], measurements of different HRV parameters [17] or subjects more than 45 years of average age [10,11,15,19].

Only one report seemed to show complicated findings [8]. In this study, cord blood mercury was associated with reduced CV_{RR}, CV_{HF}, HF, and LF in 14-year-old children, but the 7-year-old children showed only a significant association between cord blood mercury and lower LF. There are at least two possible explanations for this paradox. First, since hair mercury levels at 7 years were higher than those at 14 years (Table 1), the effect of prenatal methylmercury exposure on HRV may have been distorted by postnatal exposure. Thereafter, as postnatal exposure levels decreased with age, the influence of the prenatal exposure may have become predominant in the 14-year-old children. In support of this possibility, patients with fetal-type Minamata disease showed autonomic hypofunction approximately 45 years after the onset of the disease [7]. Second, short sleep duration in preschool children aged 5 and 6 years was associated with reduced HRV [45,46]; children who slept less than 10 h per weekday showed significantly lower CV_{RR}, CV_{LF}, CV_{HF}, LF, and HF than those who slept 10 h per weekday or more, but no significant difference in LF/HF ratio was observed between the two groups. By contrast, in 150 students aged 18–26 (mean 20) years, weekday sleep duration was not significantly associated with any HRV parameter [35]; whereas, white-collar workers with commuting times of 90 min or more showed decreased CV_{HF} compared to those with commuting time of 60–89 min or less than 60 min (2.10 ± 1.17%, 2.25 ± 1.10%, 2.58 ± 1.33%, respectively) [32]. Thus, it is possible that sleep duration may have confounded the HRV data for the 7-year-old children. Regrettably, the Faroes study did not examine the participants' sleep durations. Taken together, the data suggest that prenatal methylmercury exposure consistently affected HRV parameters in response to the exposure dose, i.e., ranging from a sympathodominant state to autonomic dysfunction. Nevertheless, HRV parameters are susceptible to physiological conditions such as sleep duration in young children [45,46] and mental stimuli [33], in addition to recent methylmercury exposure at relatively high levels.

The Seychelles child development study, which used time domain metrics, failed to find a significant association between hair mercury levels and HRV in a cohort of 19-year-olds [7]. Likewise, Faroese whaling men showed no significant associations between indicators of latent exposure to methylmercury and HRV parameters [11]. In subjects of these studies, recent exposure levels (total mercury in hair) were considerably higher (mean 9.5 μg/g for the former study and geometric mean 7.31 μg/g for the latter study) than those in the general population of other countries. Thus, comparison of HRV parameters between high/frequent and low/infrequent fish consumers should probably have been made. Autonomic function may have been affected by prenatal methylmercury exposure (though not so drastically as fetal-type Minamata disease patients [7]) and recent exposure levels, different from prenatal ones, may have changed greatly due to diversity in habits (for instance, consumption of methylmercury-contaminated fish). In two studies examining the same HRV parameters, the CV_{RR} and CV_{HF} of the above Faroese whaling men were 2.99% and 1.30%, respectively, which were lower than those of 23 Japanese healthy men aged 30–63 (mean 49) years (3.75% and 1.79%, respectively) [22]. The Seychelles child development study did not employ comparable HRV parameters. Thus, not only dose-effect relationships but also comparison between subgroups of the study population should have been attempted during the data analysis, as Varela and coworkers did [15].

In reports demonstrating a significant relationship between methylmercury exposure and HRV parameters, prenatal mercury levels showed a geometric mean 4.22 μg/g [8] and an equivalent median of 2.24 μg/g [9] in maternal hair at parturition; postnatal mercury levels were 0.83 μg/g in hair [13], 5.7 μg/L in blood [14] and 2.9 μg/L in blood [16] at the time of testing, whereas two reports except one [16] did not describe prenatal exposure levels. It would not be straightforward to estimate the critical concentration of methylmercury from these data. Valera et al. [15] and Yaginuma-Sakurai et al. [12] observed significant differences in HRV parameters between comparable subgroups with different mercury levels; specifically, the latter study indicated that a 14-week methylmercury exposure caused significant changes in some HRV parameters. Taken together, the data suggest that the average mercury level reported in the study by Yaginuma-Sakurai et al. [12] (8.76 μg/g in hair) is a reasonable estimate of a critical dose of total mercury likely to affect HRV. The EFSA panel regarded the point of departure (POD) as 11.5 μg/g for maternal mercury levels in hair and 46 μg/L for those in blood by applying a hair-to-blood ratio of 250 [5]. Since the value of 8.76 μg/g in hair corresponds to 35 μg/L in blood, the results suggest a reference dose ranging from 10 to 20 μg/L in blood after taking uncertainty factor into account; whereas, the average mercury level in whole blood of the same subjects measured after the 14-week exposure was 26.9 μg/L [47]. In such cases, intake of *n*-3 polyunsaturated fatty acids (PUFA) and selenium resulting from fish consumption should be considered in each country [12,48–50].

The effect of methylmercury on cardiac autonomic function was suggested to be reversible by an intervention study [12], separately from a cohort study by Grandjean et al. [8]. The adverse effects in the former study disappeared after 14-week wash-out period following cessation of the exposure. Recent mercury levels in hair of Faroese children aged 14 years were associated with a prolonged latency between the pons and midbrain of the auditory pathway [51], and the brainstem auditory evoked potential latency showed clear negative associations with LF and CV_{LF} [8]. Moreover, all the HRV parameters were largely reduced in comatose children with brainstem dysfunction [52], implying that impairment in the higher center of cardiac autonomic function can lead to decrease both in HF and LF power. Therefore, the pathology of autonomic imbalance due to methylmercury appears to differ for prenatal and postnatal exposures, hypothesizing that prenatal methylmercury exposure can readily impair the higher center of cardiac autonomic function and the postnatal exposure does the same in the periphery. In this case, the directionality of LF would differ, with reduced LF for prenatal exposures [8] and elevated LF for postnatal exposure [9,12,14,15].

4.2. Factors Affecting the Assessment of Cardiac Autonomic Function

The Task Force of the European Society of Cardiology and the North American Society of Pacing and Electrophysiology recommended either a 5-min recording for frequency domain metrics or a 24-h recording for time domain metrics [2]. The latter is useful for clarifying the pathophysiology of cardiac autonomic function in clinical medicine—however, whether it is useful for toxicological studies in human populations remains disputable. It would be not feasible to examine a large number of subjects with 24-h monitor, and for time domain parameters, the longer the R-R interval sample, the greater the natural variation of the signal due to heterogeneous influences on heart rate [1]. Moreover, time domain variables provide no information about the sympathetic activity [2]. None of the studies in this review used 24-h ECG monitoring. Instead, posture during measurement of R-R intervals and sampling frequency in digitizing ECG signals differed among these studies. Subject posture during measurement (supine rest, upright tilt, or sitting) affects the LF/HF ratio and the power of each component directly [2,53], and spectral signal is readily distorted by movement artifacts and ectopic beats [1]. For this reason, it is important to retain posture across measurement. Valera et al. [10,14–16], using ambulatory 2-h Holter monitoring, excluded R-R intervals whose duration was less than 80% or more than 120% of the running R-R average before performing frequency domain analysis; therefore, their data might not represent a time series of successive R-R intervals. In addition, as Table 3 shows, the data precision of R-R intervals depends on the sampling frequency [2,54]. A sampling frequency of 200 Hz or higher, in disagreement with the report by Merri et al. [54], appears to be required to preserve measurement precision. Clearly, research in human toxicology will not progress unless researchers employ comparable (and, if possible, R-R interval-adjusted) endpoints for HRV such as the CV_{RR}, CV_{HF}, and CV_{LF}.

When assessing a causal influence on HRV in humans, many potential confounders (age, sex, drinking and smoking habits, sleep duration, and mental stimuli) should be considered [3,55,56]. In the data analysis, since it is difficult to control for the effect of age in subjects at an extremely wide range of age (e.g., 5–83 years old [13]), such analyses have to be made in some age-specific groups separately. Likewise, consumption of *n*-3 PUFA such as DHA and EPA is suggested to affect HRV parameters [6,12,57], while body mass index and body fat percentage seem to be associated with corrected Q-T (QTc) intervals on ECG and heart rate, but not with HRV parameters [35]. Several disorders have been suggested to affect HRV parameters, including: Acute myocardial infarction, congestive heart failure, coronary artery disease, coronary atherosclerosis, myocardial dysfunction, cardiac transplantation, Shy-Drager syndrome, Parkinsonism, Guillain-Barre syndrome, tetraplegia, spinocerebellar degeneration, diabetic neuropathy, renal failure, chronic alcoholism, and essential hypertension [2,3]. In addition, although levels of environmental pollutants such as PCBs and lead are low in developed countries [58], concurrent exposure models would be necessary to consider the interactive effects of substances other than methylmercury [59,60]. For that reason, special attention should be paid to subjects with such disorders, as well as potential confounders and exposure to other neurotoxic substances, when conducting toxicological studies in humans.

Apart from HRV, using spectral analysis, QTc intervals are frequently used as another method for assessing autonomic nervous function [35,61–66], inasmuch as Q-T interval on ECG represents the duration between ventricular depolarization and subsequent repolarization [67]. QTc prolongation is suggested to be associated with elevated risk of heart disease and sudden cardiac death [68–71]. Therefore, the QTc interval may be a more promising indicator than HRV parameters for investigating the pathophysiology of cardiovascular events involved in methylmercury exposure, though it was not significantly associated with prenatal or postnatal methylmercury exposures at relatively low levels [9,19].

5. Conclusions

The HRV parameters analyzed using the frequency domain method, as well as CV_{RR}, appear to be more sensitive to methylmercury exposure than those using the time domain method. Most of the studies addressed in this review suggested that increased mercury levels were associated with autonomic dysfunction including a sympathodominant state, though the effect of methylmercury on HRV (e.g., LF) might differ for prenatal and postnatal exposures. In an intervention study carried out by Yaginuma-Sakurai et al., a significant difference in LF analyzed using spectral analysis was observed between the experimental group (mean mercury levels of 8.76 µg/g in hair and 26.9 µg/L in blood) and control group. Exposures near this dose may have critical effect on cardiac autonomic function. Therefore, the POD to inform a new tolerable weekly intake should be based on this critical concentration because it is lower than that established by the EFSA panel (11.5 µg/g in maternal hair). This result could probably be applicable for the general population, including pregnant women and unborn children.

Author Contributions: This publication was carried out by all the authors; K.K. conceived the idea and wrote the draft; T.I., E.M., and M.S. managed literature search and scrutinized the draft; K.M. conceived the idea, revised the draft and handled all correspondence.

Funding: This research was funded in part by a grant from the Japan Ministry of the Environment. The findings and conclusions of this article are solely the responsibility of the authors and not represent the official views of the above agency.

Conflicts of Interest: The authors declare no conflict of interest.

References

1. Bannister, R.; Mathias, C.J. (Eds.) *Autonomic Failure: A Textbook of Clinical Disorders of the Autonomic Nervous System*, 3rd ed.; Oxford University Press: Oxford, UK, 1992.
2. Camm, A.J.; Malik, M.; Bigger, J.T.; Breithardt, G.; Cerutti, S.; Cohen, R.; Coumel, P.; Fallen, E.; Kennedy, H.; Kleiger, R.E.; et al. Heart rate variability: Standards of measurement, physiological interpretation and clinical use. Task Force of the European Society of Cardiology and the North American Society of Pacing and Electrophysiology. *Circulation* **1996**, *93*, 1043–1065. [CrossRef]
3. Murata, K.; Araki, S. Assessment of autonomic neurotoxicity in occupational and environmental health as determined by ECG R–R interval variability: A review. *Am. J. Ind. Med.* **1996**, *30*, 155–163. [CrossRef]
4. Wheeler, T.; Watkins, P.J. Cardiac denervation in diabetes. *Br. Med. J.* **1973**, *8*, 584–586. [CrossRef]
5. European Food Safety Authority (EFSA). Scientific opinion on the risk for public health related to the presence of mercury and methylmercury in food. *EFSA J.* **2012**, *10*, 2985. [CrossRef]
6. Gribble, M.O.; Cheng, A.; Berger, R.D.; Rosman, L.; Guallar, E. Mercury exposure and heart rate variability: A systematic review. *Curr. Environ. Health Rep.* **2015**, *2*, 304–314. [CrossRef] [PubMed]
7. Oka, T.; Matsukura, M.; Okamoto, M.; Harada, N.; Kitano, T.; Miike, T.; Futatsuka, M. Autonomic nervous functions in fetal type Minamata disease patients: Assessment of heart rate variability. *Tohoku J. Exp. Med.* **2003**, *198*, 215–221. [CrossRef]

8. Grandjean, P.; Murata, K.; Budtz-Jørgensen, E.; Weihe, P. Cardiac autonomic activity in methylmercury neurotoxicity: 14-year follow-up of a Faroese birth cohort. *J. Pediatr.* **2004**, *144*, 169–176. [CrossRef] [PubMed]

9. Murata, K.; Sakamoto, M.; Nakai, K.; Dakeishi, M.; Iwata, T.; Liu, X.J.; Satoh, H. Subclinical effects of prenatal methylmercury exposure on cardiac autonomic function in Japanese children. *Int. Arch. Occup. Environ. Health* **2006**, *79*, 379–386. [CrossRef] [PubMed]

10. Valera, B.; Dewailly, E.; Poirier, P. Cardiac autonomic activity and blood pressure among Nunavik Inuit adults exposed to environmental mercury: A cross-sectional study. *Environ. Health* **2008**, *7*, 29. [CrossRef] [PubMed]

11. Choi, A.L.; Weihe, P.; Budtz-Jørgensen, E.; Jørgensen, P.J.; Salonen, J.T.; Tuomainen, T.P.; Murata, K.; Nielsen, H.P.; Petersen, M.S.; Askham, J.; et al. Methylmercury exposure and adverse cardiovascular effects in Faroese whaling men. *Environ. Health Perspect.* **2009**, *117*, 367–372. [CrossRef] [PubMed]

12. Yaginuma-Sakurai, K.; Murata, K.; Shimada, M.; Nakai, K.; Kurokawa, N.; Kameo, S.; Satoh, H. Intervention study on cardiac autonomic nervous effects of methylmercury from seafood. *Neurotoxicol. Teratol.* **2010**, *32*, 240–245. [CrossRef] [PubMed]

13. Lim, S.; Chung, H.U.; Paek, D. Low dose mercury and heart rate variability among community residents nearby to an industrial complex in Korea. *Neurotoxicology* **2010**, *31*, 10–16. [CrossRef] [PubMed]

14. Valera, B.; Dewailly, E.; Poirier, P. Impact of mercury exposure on blood pressure and cardiac autonomic activity among Cree adults (James Bay, Quebec, Canada). *Environ. Res.* **2011**, *111*, 1265–1270. [CrossRef] [PubMed]

15. Valera, B.; Dewailly, E.; Poirier, P.; Counil, E.; Suhas, E. Influence of mercury exposure on blood pressure, resting heart rate and heart rate variability in French Polynesians: A cross-sectional study. *Environ. Health* **2011**, *10*, 99. [CrossRef] [PubMed]

16. Valera, B.; Muckle, G.; Poirier, P.; Jacobson, S.W.; Jacobson, J.L.; Dewailly, E. Cardiac autonomic activity and blood pressure among Inuit children exposed to mercury. *Neurotoxicology* **2012**, *33*, 1067–1074. [CrossRef] [PubMed]

17. Periard, D.; Begiraj, B.; Hayoz, D.; Viswanathan, B.; Evans, K.; Thurston, S.W.; Davidson, P.W.; Myers, G.J.; Bovet, P. Associations of baroreflex sensitivity, heart rate variability, and initial orthostatic hypotension with prenatal and recent postnatal methylmercury exposure in the Seychelles Child Development study at age 19 years. *Int. J. Environ. Res. Public Health* **2015**, *12*, 3395–3405. [CrossRef] [PubMed]

18. Gump, B.B.; Dykas, M.J.; MacKenzie, J.A.; Dumas, A.K.; Hruska, B.; Ewart, C.K.; Parsons, P.J.; Palmer, C.D.; Bendinskas, K. Background lead and mercury exposures: Psychological and behavioral problems in children. *Environ. Res.* **2017**, *158*, 576–582. [CrossRef] [PubMed]

19. Miller, C.; Karimi, R.; Silbernagel, S.; Kostrubiak, D.; Schiavone, F.; Zhang, Q.; Yang, J.; Rashba, E.; Meliker, J.R. Mercury, omega-3 fatty acids, and seafood intake are not associated with heart rate variability or QT interval. *Arch. Environ. Occup. Health* **2018**, *73*, 251–257. [CrossRef] [PubMed]

20. Goldberger, A.L.; Stein, P.K. Evaluation of Heart Rate Variability. 2018. Available online: https://www.uptodate.com/contents/evaluation-of-heart-rate-variability (accessed on 20 March 2017).

21. Malliani, A.; Pagani, M.; Lombardi, F.; Cerutti, S. Cardiovascular neural regulation explored in the frequency domain. *Circulation* **1991**, *84*, 482–492. [CrossRef] [PubMed]

22. Murata, K.; Araki, S.; Yokoyama, K.; Sata, F.; Yamashita, K.; Ono, Y. Autonomic neurotoxicity of alcohol assessed by heart rate variability. *J. Auton. Nerv. Syst.* **1994**, *48*, 105–111. [CrossRef]

23. Murata, K.; Araki, S. Autonomic nervous system dysfunction in workers exposed to lead, zinc, and copper in relation to peripheral nerve conduction: A study of R–R interval variability. *Am. J. Ind. Med.* **1991**, *20*, 663–671. [CrossRef] [PubMed]

24. Murata, K.; Araki, S.; Maeda, K. Autonomic and peripheral nervous system dysfunction in workers exposed to hand-arm vibration: A study of R–R interval variability and distribution of nerve conduction velocities. *Int. Arch. Occup. Environ. Health* **1991**, *63*, 205–211. [CrossRef] [PubMed]

25. Murata, K.; Araki, S.; Yokoyama, K.; Maeda, K. Autonomic and peripheral nervous system dysfunction in workers exposed to mixed organic solvent. *Int. Arch. Occup. Environ. Health* **1991**, *63*, 335–340. [CrossRef] [PubMed]

26. Murata, K.; Araki, S.; Yokoyama, K. Assessment of the peripheral, central and autonomic nervous system function in styrene workers. *Am. J. Ind. Med.* **1991**, *20*, 775–784. [CrossRef] [PubMed]

27. Murata, K.; Araki, S.; Yokoyama, K.; Tanigawa, T.; Yamashita, K.; Okajima, F.; Sakai, T.; Matsunaga, C.; Suwa, K. Cardiac autonomic dysfunction in rotogravure printers exposed to toluene in relation to peripheral nerve conduction. *Ind. Health* **1993**, *31*, 79–90. [CrossRef] [PubMed]

28. Murata, K.; Araki, S.; Yokoyama, K.; Yamashita, K.; Okajima, F.; Nakaaki, K. Changes in autonomic function as determined by ECG R–R interval variability in sandal, shoe and leather workers exposed to *n*-hexane, xylene, and toluene. *Neurotoxicology* **1994**, *15*, 867–876. [PubMed]

29. Murata, K.; Araki, S.; Yokoyama, K.; Nomiyama, K.; Nomiyama, H.; Tao, Y.X.; Liu, S.J. Autonomic and central nervous system effects of lead in female glass workers in China. *Am. J. Ind. Med.* **1995**, *28*, 233–244. [CrossRef] [PubMed]

30. Murata, K.; Araki, S.; Okajima, F.; Nakao, M.; Suwa, K.; Matsunaga, C. Effects of occupational use of vibrating tools in the autonomic, central and peripheral nervous system. *Int. Arch. Occup. Environ. Health* **1997**, *70*, 94–100. [CrossRef] [PubMed]

31. Murata, K.; Araki, S.; Yokoyama, K.; Okumura, T.; Ishimatsu, S.; Takasu, N.; White, R.F. Asymptomatic sequelae to acute sarin poisoning in the central and autonomic nervous system 6 months after the Tokyo subway attack. *J. Neurol.* **1997**, *244*, 601–606. [CrossRef] [PubMed]

32. Kageyama, T.; Nishikido, N.; Kobayashi, T.; Kurokawa, Y.; Kabuto, M. Commuting, overtime, and cardiac autonomic activity in Tokyo. *Lancet* **1997**, *350*, 639. [CrossRef]

33. Pagani, M.; Lucini, D.; Mela, G.S.; Langewitz, W.; Malliani, A. Sympathetic overactivity in subjects complaining of unexplained fatigue. *Clin. Sci.* **1994**, *87*, 655–661. [CrossRef] [PubMed]

34. Akagi, H.; Grandjean, P.; Takizawa, T.; Weihe, P. Methylmercury dose estimation from umbilical cord concentrations in patients with Minamata disease. *Environ. Res.* **1998**, *77*, 98–103. [CrossRef] [PubMed]

35. Arai, K.; Nakagawa, Y.; Iwata, T.; Horiguchi, H.; Murata, K. Relationships between QT interval and heart rate variability at rest and the covariates in healthy young adults. *Auton. Neurosci.* **2013**, *173*, 53–57. [CrossRef] [PubMed]

36. Pagani, M.; Lombardi, F.; Guzzetti, S.; Rimoldi, O.; Furlan, R.; Pizzinelli, P.; Sandrome, G.; Malfatto, G.; Dell'Orto, S.; Piccaluga, E.; et al. Power spectral analysis of heart rate and arterial pressure variabilities as a marker of sympatho-vagal interaction in man and conscious dog. *Circulation Res.* **1986**, *59*, 178–193. [CrossRef] [PubMed]

37. Billman, G.E. The LF/HF ratio does not accurately measure cardiac sympatho-vagal balance. *Front. Physiol.* **2013**, *4*, 26. [CrossRef] [PubMed]

38. Harada, M. Minamata disease: Methylmercury poisoning in Japan caused by environmental pollution. *Crit. Rev. Toxicol.* **1995**, *25*, 1–24. [CrossRef] [PubMed]

39. Grandjean, P.; Weihe, P.; White, R.F.; Debes, F.; Araki, S.; Yokoyama, K.; Murata, K.; Sørensen, N.; Dahl, R. Cognitive deficit in 7-year-old children with prenatal exposure to methylmercury. *Neurotoxicol. Teratol.* **1997**, *19*, 417–428. [CrossRef]

40. Debes, F.; Budtz-Jørgensen, E.; Weihe, P.; White, R.F.; Grandjean, P. Impact of prenatal methylmercury exposure on neurobehavioral function at age 14 years. *Neurotoxicol. Teratol.* **2006**, *28*, 536–547. [CrossRef] [PubMed]

41. Grandjean, P.; Satoh, H.; Murata, K.; Eto, K. Adverse effects of methylmercury: Environmental health research implications. *Environ. Health Perspect.* **2010**, *118*, 1137–1145. [CrossRef] [PubMed]

42. Tatsuta, N.; Nakai, K.; Iwai-Shimada, M.; Suzuki, T.; Satoh, H.; Murata, K. Total mercury levels in hair of children aged 7 years before and after the Great East Japan earthquake. *Sci. Total Environ.* **2017**, *596–597*, 207–211. [CrossRef] [PubMed]

43. Murata, K.; Weihe, P.; Renzoni, A.; Debes, F.; Vasconcelos, R.; Zino, F.; Araki, S.; Jørgensen, P.J.; White, R.F.; Grandjean, P. Delayed evoked potentials in children exposed to methylmercury from seafood. *Neurotoxicol. Teratol.* **1999**, *21*, 343–348. [CrossRef]

44. Murata, K.; Sakamoto, M.; Nakai, K.; Weihe, P.; Dakeishi, M.; Iwata, T.; Liu, X.J.; Ohno, T.; Kurosawa, T.; Kamiya, K.; et al. Effects of methylmercury on neurodevelopment in Japanese children in relation to the Madeiran study. *Int. Arch. Occup. Environ. Health* **2004**, *77*, 571–579. [CrossRef] [PubMed]

45. Sampei, M.; Murata, K.; Dakeishi, M.; Wood, D.C. Cardiac autonomic hypofunction in preschool children with short nocturnal sleep. *Tohoku J. Exp. Med.* **2006**, *208*, 235–242. [CrossRef] [PubMed]

46. Sampei, M.; Dakeishi, M.; Wood, D.C.; Iwata, T.; Murata, K. Spontaneous awakening from nocturnal sleep and cardiac autonomic function in preschool children. *Auton. Neurosci.* **2007**, *133*, 170–174. [CrossRef] [PubMed]

47. Yaginuma-Sakurai, K.; Murata, K.; Iwai-Shimada, M.; Nakai, K.; Kurokawa, N.; Tatsuta, N.; Satoh, H. Hair-to-blood ratio and biological half-life of mercury: Experimental study of methylmercury exposure through fish consumption in humans. *J. Toxicol. Sci.* **2012**, *37*, 123–130. [CrossRef] [PubMed]

48. Mahaffey, K.R.; Clickner, R.P.; Jeffries, R.A. Methylmercury and omega-3 fatty acids: Co-occurrence of dietary sources with emphasis on fish and shellfish. *Environ. Res.* **2008**, *107*, 20–29. [CrossRef] [PubMed]

49. Strain, J.J.; Davidson, P.W.; Bonham, M.P.; Duffy, E.M.; Stokes-Riner, A.; Thurston, S.W.; Wallace, J.M.; Robson, P.J.; Shamlaye, C.F.; Georger, L.A.; et al. Associations of maternal long-chain polyunsaturated fatty acids, methylmercury, and infant development in the Seychelles Child Development Nutrition study. *Neurotoxicology* **2008**, *29*, 776–782. [CrossRef] [PubMed]

50. Nakamura, M.; Hachiya, N.; Murata, K.Y.; Nakanishi, I.; Kondo, T.; Yasutake, A.; Miyamoto, K.; Ser, P.H.; Omi, S.; Furusawa, H.; et al. Methylmercury exposure and neurological outcomes in Taiji residents accustomed to consuming whale meat. *Environ. Int.* **2014**, *68*, 25–32. [CrossRef] [PubMed]

51. Murata, K.; Weihe, P.; Budtz-Jørgensen, E.; Jørgensen, P.J.; Grandjean, P. Delayed brainstem auditory evoked potential latencies in 14-year-old children exposed to methylmercury. *J. Pediatr.* **2004**, *144*, 177–183. [CrossRef] [PubMed]

52. Shimomura, C.; Matsuzaka, T.; Koide, E.; Kinoshita, S.; Ono, Y.; Tsuji, Y.; Kawasaki, C.; Suzuki, Y. Spectral analysis of heart rate variability in the dysfunction of brainstem. *Brain Dev.* **1991**, *23*, 26–31. (In Japanese)

53. Hayano, J.; Mukai, S.; Sakakibara, M.; Okada, A.; Takata, K.; Fujinami, T. Effects of respiratory interval on vagal modulation of heart rate. *Am. J. Physiol.* **1994**, *267*, H33–H40. [CrossRef] [PubMed]

54. Merri, M.; Farden, D.C.; Mottley, J.G.; Titlebaum, E.L. Sampling frequency of the electrocardiogram for spectral analysis of the heart rate variability. *IEEE Trans. Biomed. Eng.* **1990**, *37*, 99–106. [CrossRef] [PubMed]

55. Murata, K.; Landrigan, P.J.; Araki, S. Effects of age, heart rate, gender, tobacco and alcohol ingestion on R–R interval variability in human ECG. *J. Auton. Nerv. Syst.* **1992**, *37*, 199–206. [CrossRef]

56. Koenig, J.; Thayer, J.F. Sex differences in healthy human heart rate variability: A meta-analysis. *Neurosci. Biobehav. Rev.* **2016**, *64*, 288–310. [CrossRef] [PubMed]

57. Valera, B.; Suhas, E.; Counil, E.; Poirier, P.; Dewailly, E. Influence of polyunsaturated fatty acids on blood pressure, resting heart rate and heart rate variability among French Polynesians. *J. Am. Coll. Nutr.* **2014**, *33*, 288–296. [CrossRef] [PubMed]

58. Konishi, Y.; Kuwabara, K.; Hori, S. Continuous survellance of organochlorine compounds in human breast milk from 1972 to 1998 in Osaka, Japan. *Arch. Environ. Contam. Toxicol.* **2001**, *40*, 571–578. [CrossRef] [PubMed]

59. Yorifuji, T.; Debes, F.; Weihe, P.; Grandjean, P. Prenatal exposure to lead and cognitive deficit in 7- and 14-year-old children in the presence of concomitant exposure to similar molar concentration of methylmercury. *Neurotoxicol. Teratol.* **2011**, *33*, 205–211. [CrossRef] [PubMed]

60. Tatsuta, N.; Kurokawa, N.; Nakai, K.; Suzuki, K.; Iwai-Shimada, M.; Murata, K.; Satoh, H. Effects of intrauterine exposures to polychlorinated biphenyls, methylmercury, and lead on birth weight in Japanese male and female newborn. *Environ. Health Prev. Med.* **2017**, *22*, 39. [CrossRef] [PubMed]

61. Murata, K.; Yano, E.; Shinozaki, T. Impact of shift work on cardiovascular functions in a 10-year follow-up study. *Scand. J. Work Environ. Health* **1999**, *25*, 272–277. [CrossRef] [PubMed]

62. Murata, K.; Yano, E.; Shinozaki, T. Cardiovascular dysfunction due to shift work. *J. Occup. Enviorn. Med.* **1999**, *41*, 748–753. [CrossRef]

63. Ishii, N.; Dakeishi, M.; Sasaki, M.; Iwata, T.; Murata, K. Cardiac autonomic imbalance in female nurses with shift work. *Auton. Neurosci.* **2005**, *122*, 94–99. [CrossRef] [PubMed]

64. Rautaharju, P.M.; Surawicz, B.; Gettes, L.S.; Bailey, J.J.; Childers, R.; Deal, B.J.; Gorgels, A.; Hancock, E.W.; Josephson, M.; Kligfield, P.; et al. AHA/ACCF/HRS recommendations for the standardization and interpretation of the electrocardiogram: Part IV: The ST segment, T and U waves, and the QT interval: A scientific statement from the American Heart Association Electrocardiography and Arrhythmias Committee, Council on Clinical Cardiology; the American College of Cardiology Foundation; and the Heart Rhythm Society: Endorsed by the International Society for Computerized Electrocardiology. *Circulation* **2009**, *119*, e241–e250. [CrossRef] [PubMed]

65. Maeda, E.; Iwata, T.; Murata, K. Effect of work stress and home stress on autonomic nervous function in Japanese male workers. *Ind. Health* **2015**, *53*, 132–138. [CrossRef] [PubMed]

66. Enoki, M.; Maeda, E.; Iwata, T.; Murata, K. The association between work-related stress and autonomic imbalance among call center employees in Japan. *Tohoku J. Exp. Med.* **2017**, *243*, 321–328. [CrossRef] [PubMed]

67. Yun, J.; Hwangbo, E.; Lee, J.; Chon, C.-R.; Kim, P.A.; Joeong, I.-H.; Park, M.; Park, R.; Kang, S.-J.; Choi, D. Analysis of an ECG record database reveals QT interval prolongation potential of famotidine in a large Korean population. *Cardiovasc. Toxicol.* **2015**, *15*, 197–202. [CrossRef] [PubMed]

68. Dekker, J.M.; Schouten, E.G.; Klootwijk, P.; Pool, J.; Swenne, C.A.; Kromhout, D. Heart rate variability from short electrocardiographic recordings predicts mortality from all causes in middle-aged and elderly men: The Zutphen study. *Am. J. Epidemiol.* **1997**, *145*, 899–908. [CrossRef] [PubMed]

69. Montanez, A.; Ruskin, J.N.; Hebert, P.R.; Lamas, G.A.; Hennekens, C.H. Prolonged QTc interval and risks of total and cardiovascular mortality and sudden death in the general population. *Arch. Intern. Med.* **2004**, *164*, 943–948. [CrossRef] [PubMed]

70. Straus, S.M.; Kors, J.A.; de Bruin, M.L.; van der Hooft, C.S.; Kofman, A.; Heeringa, J.; Deckers, J.W.; Kingma, J.H.; Sturkenboom, M.C.J.M.; Stricker, B.H.C.; et al. Prolonged QTc interval and risk of sudden cardiac death in a population of older adults. *J. Am. Coll. Cardiol.* **2006**, *47*, 362–367. [CrossRef] [PubMed]

71. Soliman, E.Z.; Howard, G.; Cushman, M.; Kissela, B.; Kleindorfer, D.; Le, A.; Judd, S.; McClure, L.A.; Howard, V.J. Prolongation of QTc and risk of stroke: The REGARDS study. *J. Am. Coll. Cardiol.* **2012**, *59*, 1460–1467. [CrossRef] [PubMed]

toxics

MDPI

Article

Chemokine CCL4 Induced in Mouse Brain Has a Protective Role against Methylmercury Toxicity

Tsutomu Takahashi [1,2,†], Min-Seok Kim [1,3,†], Miyuki Iwai-Shimada [1,4], Masatake Fujimura [5], Takashi Toyama [1], Akira Naganuma [1] and Gi-Wook Hwang [1,*]

[1]   Laboratory of Molecular and Biochemical Toxicology, Graduate School of Pharmaceutical Sciences, Tohoku University, Aoba-ku, Sendai 980-8578, Japan; tsutomu@toyaku.ac.jp (Ts.T.); assams7@naver.com (M.-S.K.); iwai.miyuki@nies.go.jp (M.I.-S.); takashi.toyama.c6@tohoku.ac.jp (Ta.T.); naganuma@m.tohoku.ac.jp (A.N.)
[2]   Department of Environmental Health, School of Pharmacy, Tokyo University of Pharmacy and Life Sciences, 1432-1 Horinouchi, Hachioji, Tokyo 192-0392, Japan
[3]   Department of Inhalation Toxicology Research, Korea Institute of Toxicology, Jeonbuk 56212, Korea
[4]   Center for Health and Environmental Risk Research, National Institute for Environmental Studies, Onogawa 16-2, Tsukuba, Ibaraki 305-8506, Japan
[5]   Department of Basic Medical Science, National Institute for Minamata Disease, Kumamoto 867-0008, Japan; fujimura@nimd.go.jp
*   Correspondence: gwhwang@m.tohoku.ac.jp; Tel./Fax: +81-22-795-6872
†   These authors contributed equally to this work.

Received: 4 June 2018; Accepted: 5 July 2018; Published: 7 July 2018

Abstract: Methylmercury (MeHg) is selectively toxic to the central nervous system, but mechanisms related to its toxicity are poorly understood. In the present study, we identified the chemokine, C-C motif Chemokine Ligand 4 (CCL4), to be selectively upregulated in the brain of MeHg-administered mice. We then investigated the relationship between CCL4 expression and MeHg toxicity using in vivo and in vitro approaches. We confirmed that in C17.2 cells (a mouse neural stem cell line) and the mouse brain, induction of CCL4 expression occurs prior to cytotoxicity caused by MeHg. We also show that the addition of recombinant CCL4 to the culture medium of mouse primary neurons attenuated MeHg toxicity, while knockdown of CCL4 in C17.2 cells resulted in higher MeHg sensitivity compared with control cells. These results suggest that CCL4 is a protective factor against MeHg toxicity and that induction of CCL4 expression is not a result of cytotoxicity by MeHg but is a protective response against MeHg exposure.

Keywords: methylmercury; brain; chemokine; CCL4

---

1. Introduction

Methylmercury (MeHg) is an environmental pollutant well known to cause Minamata disease [1]. MeHg causes central nervous system (CNS) disorders whose main symptoms include sensory paralysis, speech disorder, ataxia, and visual field narrowing [2,3]. MeHg can easily pass through the blood-placental barrier and can therefore affect the brain of the immature fetus [4,5]. Although MeHg is selectively toxic in the brain, mechanisms related to this selectivity and defense mechanisms are still unknown. We have analyzed patterns of gene expression in the brains of mice that were administered MeHg and have identified a number of upregulated genes [6–8]. Among these genes, many encode cytokines, such as chemokines and interleukins. One such gene is C-C motif Chemokine Ligand 4 (CCL4), which is specifically upregulated in the brain by MeHg [9]. CCL4, also called macrophage inflammatory protein 1β (MIP1β), was identified as an inflammatory protein produced from macrophages [10]. CCL4, secreted extracellularly, binds to chemokine receptors (CCR1, CCR5) and is involved in leukocyte infiltration and activation [11]. CCL4 is secreted from glial cells and astrocytes in the CNS and has been suggested to be involved in the progression of various brain diseases, including Alzheimer's disease, multiple sclerosis, and ischemic brain disease [12–15]. Nevertheless, the functions of CCL4 in the brain are unclear and the relationship between chemokines and MeHg toxicity is poorly understood. Recently, Godefroy et al. reported that CCL2, another C-C chemokine, attenuates MeHg toxicity in rat primary neuron cultures [16], but the mechanism for this is not understood. In this study, we investigated the relationship between CCL4 expression and MeHg toxicity, in vivo and in vitro, using mice and neuronal cell lines.

2. Materials and Methods

2.1. Animal Experiments

Eight-week-old male C57BL/6 mice were purchased from Japan SLC, Inc. (Shizuoka, Japan). The mice were housed in plastic cages (five animals per cage) at $22 \pm 2\ ^\circ$C with a relative humidity of $55 \pm 20\%$ under a 12-h light-dark cycle and allowed free access to chow (F-2, Oriental Yeast, Tokyo, Japan) and water. All experiments were performed in accordance with the Regulations for Animal Experiments and Related Activities at Tohoku University, and were approved by the Animal Care Committee of Tohoku University (No.: 2016PhA-001, Date: 8 February 2016). After an adaptation period, mice were randomly divided into control ($n = 5$) and MeHg-treated ($n = 5$) groups. Methylmercuric chloride (25 mg/kg), dissolved in physiological saline, was administered by subcutaneous injection. After the indicated time period, the mice were dissected and each organ was subjected to the various assays.

2.2. Immunochemistry

Immunohistochemistry was performed as described previously [17–19]. Paraffin embedded sections were cut using a microtome, and immunohistochemistry was performed using the Vectastain Elite ABC Kit (Vector Laboratories, Burlingame, CA, USA) with an antibody to neuronal nuclei (NeuN) (Chemicon, Temecula, CA, USA).

2.3. Cell Culture

Mouse C17.2 neural stem cells were cultured in Dulbecco's modified Eagle's medium (DMEM) (Nissui Pharmaceutical, Tokyo, Japan) supplemented with 10% heat-inactivated fetal bovine serum (FBS), 2 mM/L L-glutamine, and antibiotic (100 IU/mL penicillin and 100 mg/mL streptomycin) in a humidified 5% CO_2 atmosphere at 37 $^\circ$C. Mouse primary cerebellar granule cells were cultured in Neurobasal-A medium (Thermo Fisher Scientific, Waltham, MA, USA) containing 2% B25 (Thermo Fisher Scientific), 1% FBS, and 25 mM KCl in 12-well plates for 2 weeks.

2.4. siRNA Transfection

Double-stranded siRNA for CCL4 (target sequence: CTTTGTGATGGATTACTATTT) and negative control siRNA were purchased from Sigma-Aldrich (St. Louis, MO, USA). C17.2 cells were transfected with siRNAs using HiPerFect transfection reagent (Qiagen, Germantown, MD, USA) according to the manufacturer's protocol.

2.5. Cell Viability Assay

C17.2 cells and mouse primary cerebellar granule cells were cultured in media containing methylmercuric chloride for 24 h. Cell viability was measured using the alamarBlue® assay (Biosource, Camarillo, CA, USA). Fluorescence was measured using a Gemini XPS microplate spectrofluorometer (Molecular Devices, Sunnyvale, CA, USA) (excitation wavelength 545 nm; emission wavelength 590 nm). Trypan blue assays were performed using a Vi-Cell XR cell viability analyzer (Beckman coulter, San Diego, CA, USA).

2.6. Measurement of CCL4 mRNA Levels by Quantitative Real-Time PCR

Total RNA from organs and cells was isolated using the Isogen II Kit (Nippon Gene, Tokyo, Japan) according to the manufacturer's protocol. The first-strand cDNA was synthesized from 500 ng of total RNA using the PrimeScriptTM RT Reagent Kit (Takara, Shiga, Japan). Quantitative real-time PCR analysis was performed using SYBR Premix EX Taq (Takara) with a Thermal Cycler Dice® (Takara). The PCR primers used included the following: CCL4, 5′-ACCCTGTGACATTTCACGGAG-3′ (sense) and 5′-GTACTCGATTGATAGAGGAC-3′ (antisense); and GAPDH, 5′-ATCACCATCTTCCAGGAGCGA-3′ (sense) and 5′-AGGGGCCATCCACAGTCTT-3′ (antisense). Fold changes in mRNA levels were determined from standard curves after calibration of the assay. CCL4 mRNA levels were normalized to those of GAPDH.

2.7. Statistical Analysis

If not stated otherwise, statistical significance of the data was determined using analysis of variance (ANOVA) with Dunnett's post hoc test.

3. Results

3.1. CCL4 Expression Is Induced Prior to Neuronal Damage Caused by MeHg

To study the relationship between MeHg toxicity and CCL4 expression in mice, we administered a single dose of methylmercuric chloride (25 mg/kg) and then used real-time qPCR to investigate changes in CCL4 mRNA levels in cerebrum, cerebellum, kidney, and liver over time (Figure 1A). CCL4 mRNA levels were elevated in the cerebrum and cerebellum from 5 days after MeHg administration (Figure 1B). CCL4 expression was not induced in the kidney or liver at any time point tested (Figure 1B). This indicates that MeHg induces CCL4 expression in a brain-specific manner. We then investigated pathological changes in the brains of mice after single-dose administration of 25 mg/kg methylmercuric chloride. As a reference index, we counted cells that were positive for the neuronal marker, NeuN. We observed almost no change in NeuN-positive cell numbers in the cerebellum, even at 7 days after MeHg administration (data not shown). In the cerebrum, no changes were found in the number of NeuN-positive cells up to 5 days, and on day 7 a slight decrease in the number of NeuN-positive cells was observed (Figure 1C). Therefore, CCL4 expression was induced in the mouse brain prior to neuronal damage caused by MeHg.

3.2. CCL4 Attenuates MeHg Toxicity in Primary Mouse Neuron Cultures

Godefroy et al. reported that addition of recombinant CCL2 into the medium of primary rat neuron cultures attenuated MeHg toxicity [16]. Therefore, we used primary mouse neuron cultures to investigate the effect of chemokines on MeHg sensitivity. Addition of recombinant CCL2 to the culture medium significantly attenuated the MeHg toxicity on primary mouse neurons (Figure 2A). The addition of recombinant CCL4 also significantly attenuated MeHg toxicity, to a greater extent than CCL2, indicating that CCL4, like CCL2, is a protective factor against MeHg neurotoxicity (Figure 2B).

3.3. CCL4 Expression Is Induced Prior to MeHg-Induced Cytotoxicity in C17.2 Cells

We investigated the effect of MeHg on CCL4 expression in C17.2 cells, a neural progenitor cell derived from the mouse brain. CCL4 expression was below detection threshold under normal conditions. However, increased CCL4 mRNA levels were detected from 2 h after MeHg treatment, and this increase continued over time up to 9 h (Figure 3A). However, when cell viability was measured by the trypan blue assay, cell viability decreased from 9 h after MeHg treatment (Figure 3B). This showed that even in C17.2 cells, CCL4 expression is induced prior to MeHg cytotoxicity.

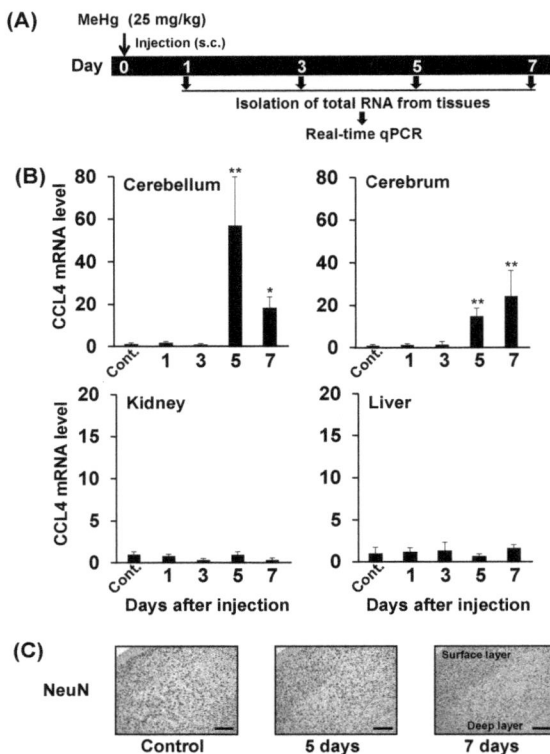

Figure 1. Relationship between neurological damage and CCL4 expression in the brains of mice treated with methylmercury. C57BL/6 mice were injected once subcutaneously with methylmercuric chloride (MeHg) (25 mg/kg). (**A**) Selected organs (cerebellum, cerebrum, liver and kidney) were dissected 1, 3, 5, or 7 days after the injection. Cont.: Control. (**B**) CCL4 mRNA levels in each organ were measured by quantitative real-time PCR. Data are represented as mean $\pm$ SD, * $p < 0.05$; ** $p < 0.01$ compared with "Control". (**C**) NeuN (neuron marker)-positive cells in the cerebral cortex detected by immunostaining. Scale bars represent 250 μm.

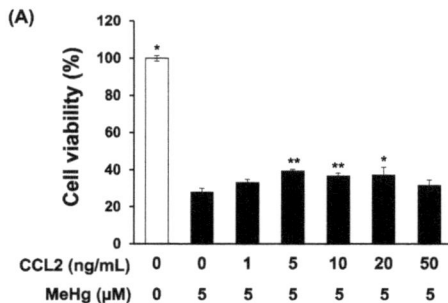

Figure 2. *Cont.*

(B)

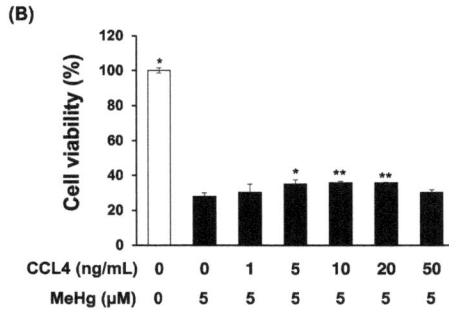

Figure 2. Effects of recombinant CCL2 or CCL4 on methylmercury-induced cytotoxicity in mouse primary cerebellar granule cells. Mouse primary cerebellar granule cells (neurons) (1×10^6 cells/mL) were cultured in Neurobasal A medium containing 2% B25, 1% FBS, and 25 mM KCl in 12-well plates for 2 weeks. Recombinant CCL2 (**A**) or CCL4 (**B**) was then added to the culture medium, and 1 h later cells were exposed to 5 μM methylmercuric chloride (MeHg) for 24 h. Cell viability was measured by the alamarBlue® assay. Data are presented as the mean ± S.D. * $p < 0.05$; ** $p < 0.01$ compared with the "0 ng/mL recombinant chemokine, 5 μM MeHg group".

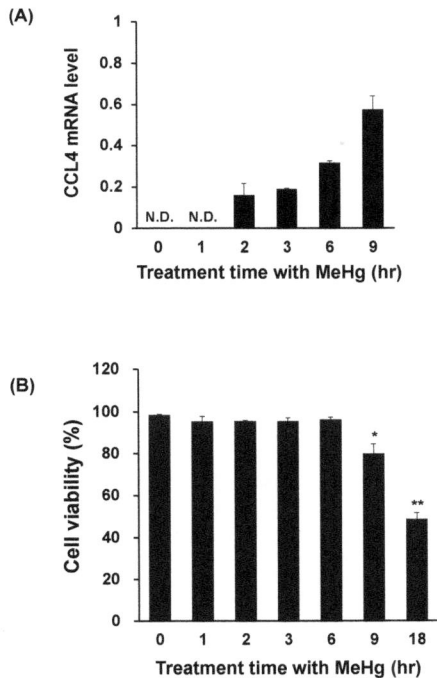

(A)

(B)

Figure 3. Relationship between cell viability and CCL4 expression in methylmercury-treated mouse C17.2 neural stem cells. C17.2 cells (4×10^5 cells/2 mL) were seeded into 6-well plates. After incubation for 18 h, cells were treated with 10 μM methylmercuric chloride (MeHg) for the indicated times. (**A**) CCL4 mRNA levels were measured by quantitative real-time PCR. (**B**) Cell viability was determined by the trypan blue assay using the Vi-CELL cell counter. N.D.: not detected. Data are presented as the mean ± S.D. * $p < 0.05$; ** $p < 0.01$ compared with "0 h group".

3.4. Knockdown of CCL4 Enhances MeHg Cytotoxicity in C17.2 Cells

To clarify the relationship between CCL4 expression and MeHg toxicity, we investigated the effects of CCL4 knockdown on the MeHg sensitivity of C17.2 cells. The introduction of siRNA to CCL4 reduced the induction of CCL4 expression by MeHg by about 60% (Figure 4A). CCL4 knockdown cells were more sensitive to MeHg than cells transfected with control siRNA (Figure 4B). This indicates that CCL4 has a protective role against MeHg toxicity.

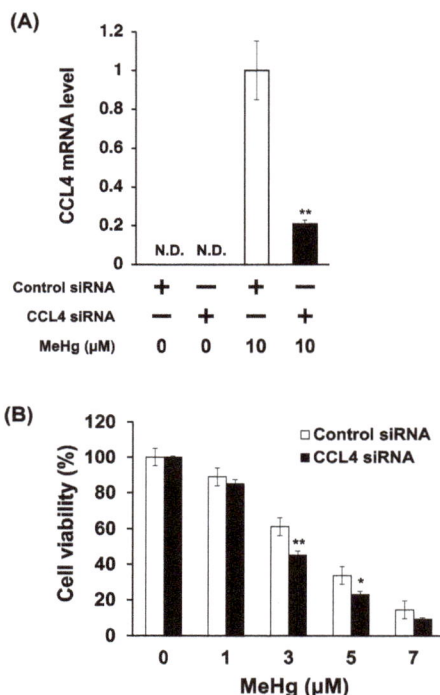

Figure 4. Effects of CCL4 knockdown on the sensitivity of C17.2 cells to methylmercury. C17.2 cells (1×10^4 cells/90 μL) transfected with CCL4 siRNA were seeded into 96-well plates. After incubation for 18 h, transfected cells were treated with indicated concentrations of methylmercuric chloride (MeHg) for 24 h. (**A**) CCL4 mRNA levels were measured by quantitative real-time PCR. (**B**) Cell viability was measured by the alamarBlue® assay. Data are presented as the mean ± S.D. N.D.: not detected. Statistical significance when compared to corresponding "control siRNA group" at each concentration point: * $p < 0.05$, ** $p < 0.01$.

4. Discussion

The present study showed that CCL4 expression in mouse-derived neuronal precursor C17.2 cells, as well as in the mouse brain, was induced prior to MeHg cytotoxicity. These findings indicate the possibility that the CCL4 expression induced by MeHg is not the result of cellular damage caused by MeHg, but a protective response to MeHg exposure. It is also known that the cell has a number of protective systems against toxicity of environmental toxicants and that those are activated in the early stage of exposure. Therefore, CCL4 induction in the early stage of MeHg exposure seems to be a protective action against its toxicity.

The transcription factor, NF-κB, is involved in the induction of cytokine expression [20,21]. CCL4 expression is induced in macrophages by lipopolysaccharide (LPS) and hydrogen peroxide and NF-κB is involved in this induction [22–24]. We recently found that knockdown of p65, a subunit of NF-κB, slightly suppressed the induction of CCL4 expression in response to MeHg in C17.2 cells (unpublished data). This suggests that while NF-κB is partially involved in the induction of CCL4 expression in response to MeHg, other transcription factors are also involved. It is possible that MeHg induces CCL4 expression due to the activation of transcription factors that differ from those activated by LPS and hydrogen peroxide. In future studies, we hope to clarify the mechanisms underlying brain-specific MeHg toxicity by determining how CCL4 is induced in response to MeHg.

In this study, CCL4 was newly identified as a protective factor against MeHg induced cytotoxicity. Recently, CCL2 was also identified as a protective factor against MeHg toxicity on primary rat neurons [16]. Although the protective mechanisms against MeHg toxicity through CCL2 and CCL4 are unknown, it is reported that expressions of CCL2 and CCL4 are increased early in the cerebral tissue of patients with posttraumatic brain contusions [25]. This suggests that both chemokines may play an important role as a defensive response in the brain injury.

It is known that CCL4 induces the production of cytokines including interleukin (IL)-1 and IL-6 [26–28]. Recently, Noguchi et al. reported that IL-6 expression is induced by MeHg, and that IL-6 has a protective role against MeHg-induced neurotoxicity [29]. It is thus thought that increased levels of CCL4, in response to MeHg, may enhance IL-6 production, thereby reducing MeHg toxicity. However, we could not confirm the induction of IL-6 expression in mouse brains treated with MeHg (data not shown). This suggests that IL-6 is not involved in reducing MeHg toxicity by CCL4.

MeHg induced higher CCL4 expression in the cerebellum compared with the cerebrum (Figure 1B). Nevertheless, loss of neurons was only observed in the cerebrum and was not confirmed in the cerebellum (Figure 1C). We have reported that tumor necrosis factor-α (TNF-α), an inflammatory cytokine, is selectively induced in the brain of mice treated with MeHg, and that the degree of induction was larger in the cerebrum than in the cerebellum [8]. In addition, TNF-α may enhance MeHg toxicity because a TNF-α antagonist attenuated MeHg toxicity for C17.2 cells [8]. Based on these findings, the degree of induction of CCL4 expression, which reduces MeHg toxicity, was higher in the cerebellum than in the cerebrum, and the degree of induction of TNF-α expression, which enhances MeHg toxicity, was low in the cerebellum. Therefore, MeHg toxicity may not be observed in the cerebellum under our experimental conditions. In addition, we found that the expression of CCL3 and of IL-19 are also specifically induced in the brain by MeHg [9,30]. Therefore, there are many cytokines that are specifically induced in the mouse brain by MeHg, and the combined action of these may play an important role in neuronal damage caused by MeHg. In future studies, we hope to clarify crosstalk between cytokine molecules that are specifically induced in the brain, which will help to elucidate the mechanisms involved in brain-specific MeHg toxicity.

Author Contributions: T.T. (Tsutomu Takahashi), A.N. and G.-W.H. designed the experiments and wrote the manuscript. M.-S.K. and T.T. (Tsutomu Takahashi) prepared Figures 1 and 2. M.F. prepared Figure 1C. M.-S.K. and M.I.-S. prepared Figure 3. T.T. (Tsutomu Takahashi) and T.T. (Takashi Toyama) prepared Figure 4.

Funding: This work was supported by a Grant-in-Aid for Scientific Research from the Japanese Society for the Promotion of Science (#15H05714, #16H02961).

Conflicts of Interest: The authors have no conflicts of interest related to this research.

References

1. Harada, M. Minamata disease: Methylmercury poisoning in Japan caused by environmental pollution. *Crit. Rev. Toxicol.* **1995**, *25*, 1–24. [CrossRef] [PubMed]
2. Aschner, M.; Aschner, J.L. Mercury neurotoxicity: Mechanisms of blood-brain barrier transport. *Neurosci. Biobehav. Rev.* **1990**, *14*, 169–176. [CrossRef]

3. Vahter, M.; Akesson, A.; Lind, B.; Bjors, U.; Schutz, A.; Berglund, M. Longitudinal study of methylmercury and inorganic mercury in blood and urine of pregnant and lactating women, as well as in umbilical cord blood. *Environ. Res.* **2000**, *84*, 186–194. [CrossRef] [PubMed]

4. Ballatori, N. Transport of toxic metals by molecular mimicry. *Environ. Health. Perspect.* **2002**, *110*, 689–694. [CrossRef] [PubMed]

5. Castoldi, A.F.; Johansson, C.; Onishchenko, N.; Coccini, T.; Roda, E.; Vahter, M.; Ceccatelli, S.; Manzo, L. Human developmental neurotoxicity of methylmercury: Impact of variables and risk modifiers. *Regul. Toxicol. Pharmacol.* **2008**, *51*, 201–214. [CrossRef] [PubMed]

6. Hwang, G.W.; Lee, J.Y.; Ryoke, K.; Matsuyama, F.; Kim, J.M.; Takahashi, T.; Naganuma, A. Gene expression profiling using DNA microarray analysis of the cerebellum of mice treated with methylmercury. *J. Toxicol. Sci.* **2011**, *36*, 389–391. [CrossRef] [PubMed]

7. Lee, J.Y.; Hwang, G.W.; Kim, M.S.; Takahashi, T.; Naganuma, A. Methylmercury induces a brain-specific increase in chemokine CCL4 expression in mice. *J. Toxicol. Sci.* **2012**, *37*, 1279–1282. [CrossRef] [PubMed]

8. Iwai-Shimada, M.; Takahashi, T.; Kim, M.S.; Fujimura, M.; Ito, H.; Toyama, T.; Naganuma, A.; Hwang, G.W. Methylmercury induces the expression of TNF-α selectively in the brain of mice. *Sci. Rep.* **2016**, *6*, 38294. [CrossRef] [PubMed]

9. Kim, M.S.; Takahashi, T.; Lee, J.Y.; Hwang, G.W.; Naganuma, A. Global chemokine expression in methylmercury-treated mice: Methylmercury induces brain-specific expression of CCL3 and CCL4. *J. Toxicol. Sci.* **2013**, *38*, 925–929. [CrossRef] [PubMed]

10. Maurer, M.; von Stebut, E. Macrophage inflammatory protein-1. *Int. J. Biochem. Cell Biol.* **2004**, *36*, 1882–1886. [CrossRef] [PubMed]

11. Ren, M.; Guo, Q.; Guo, L.; Lenz, M.; Qian, F.; Koenen, R.R.; Xu, H.; Schilling, A.B.; Weber, C.; Ye, R.D.; et al. Polymerization of MIP-1 chemokine (CCL3 and CCL4) and clearance of MIP-1 by insulin-degrading enzyme. *EMBO J.* **2010**, *29*, 3952–3966. [CrossRef] [PubMed]

12. Cowell, R.M.; Xu, H.; Galasso, J.M.; Silverstein, F.S. Hypoxic-ischemic injury induces macrophage inflammatory protein-1α expression in immature rat brain. *Stroke* **2002**, *33*, 795–801. [CrossRef] [PubMed]

13. Boven, L.A.; Montagne, L.; Nottet, H.S.; De Groot, C.J. Macrophage inflammatory protein-1α (MIP-1α), MIP-1β, and RANTES mRNA semiquantification and protein expression in active demyelinating multiple sclerosis (MS) lesions. *Clin. Exp. Immunol.* **2000**, *122*, 257–263. [CrossRef] [PubMed]

14. Xia, M.Q.; Qin, S.X.; Wu, L.J.; Mackay, C.R.; Hyman, B.T. Immunohistochemical study of the β-chemokine receptors CCR3 and CCR5 and their ligands in normal and Alzheimer's disease brains. *Am. J. Pathol.* **1998**, *153*, 31–37. [CrossRef]

15. Szczucinski, A.; Losy, J. Chemokines and chemokine receptors in multiple sclerosis. Potential targets for new therapies. *Acta. Neurol. Scand.* **2007**, *115*, 137–146. [CrossRef] [PubMed]

16. Godefroy, D.; Gosselin, R.D.; Yasutake, A.; Fujimura, M.; Combadiere, C.; Maury-Brachet, R.; Laclau, M.; Rakwal, R.; Melik-Parsadaniantz, S.; Bourdineaud, J.P.; et al. The chemokine CCL2 protects against methylmercury neurotoxicity. *Toxicol. Sci.* **2012**, *125*, 209–218. [CrossRef] [PubMed]

17. Fujimura, M.; Usuki, F. Methylmercury causes neuronal cell death through the suppression of the TrkA pathway: In vitro and in vivo effects of TrkA pathway activators. *Toxicol. Appl. Pharmacol.* **2015**, *282*, 259–266. [CrossRef] [PubMed]

18. Fujimura, M.; Usuki, F. In situ different antioxidative systems contribute to the site-specific methylmercury neurotoxicity in mice. *Toxicology* **2017**, *392*, 55–63. [CrossRef] [PubMed]

19. Fujimura, M.; Usuki, F. Site-specific neural hyperactivity via the activation of MAPK and PKA/CREB pathways triggers neuronal degeneration in methylmercury-intoxicated mice. *Toxicol. Lett.* **2017**, *271*, 66–73. [CrossRef] [PubMed]

20. Thompson, W.L.; Van Eldik, L.J. Inflammatory cytokines stimulate the chemokines CCL2/MCP-1 and CCL7/MCP-3 through NFkB and MAPK dependent pathways in rat astrocytes [corrected]. *Brain Res.* **2009**, *1287*, 47–57. [CrossRef] [PubMed]

21. Giraud, S.N.; Caron, C.M.; Pham-Dinh, D.; Kitabgi, P.; Nicot, A.B. Estradiol inhibits ongoing autoimmune neuroinflammation and NFkappaB-dependent CCL2 expression in reactive astrocytes. *Proc. Natl. Acad. Sci. USA* **2010**, *107*, 8416–8421. [CrossRef] [PubMed]

22. Proffitt, J.; Crabtree, G.; Grove, M.; Daubersies, P.; Bailleul, B.; Wright, E.; Plumb, M. An ATF/CREB-binding site is essential for cell-specific and inducible transcription of the murine *MIP-1β* cytokine gene. *Gene* **1995**, *152*, 173–179. [CrossRef]

23. Wiesner, P.; Choi, S.H.; Almazan, F.; Benner, C.; Huang, W.; Diehl, C.J.; Gonen, A.; Butler, S.; Witztum, J.L.; Glass, C.K.; et al. Low doses of lipopolysaccharide and minimally oxidized low-density lipoprotein cooperatively activate macrophages via nuclear factor kappa B and activator protein-1: Possible mechanism for acceleration of atherosclerosis by subclinical endotoxemia. *Circ. Res.* **2010**, *107*, 56–65. [CrossRef] [PubMed]

24. Shi, M.M.; Godleski, J.J.; Paulauskis, J.D. Regulation of macrophage inflammatory protein-1α mRNA by oxidative stress. *J. Biol. Chem.* **1996**, *271*, 5878–5883. [CrossRef] [PubMed]

25. Stefini, R.; Catenacci, E.; Piva, S.; Sozzani, S.; Valerio, A.; Bergomi, R.; Cenzato, M.; Mortini, P.; Latronico, N. Chemokine detection in the cerebral tissue of patients with posttraumatic brain contusions. *J. Neurosurg.* **2008**, *108*, 958–962. [CrossRef] [PubMed]

26. Bless, N.M.; Huber-Lang, M.; Guo, R.F.; Warner, R.L.; Schmal, H.; Czermak, B.J.; Shanley, T.P.; Crouch, L.D.; Lentsch, A.B.; Sarma, V.; et al. Role of CC chemokines (macrophage inflammatory protein-1β, monocyte chemoattractant protein-1, RANTES) in acute lung injury in rats. *J. Immunol.* **2000**, *164*, 2650–2659. [CrossRef] [PubMed]

27. Fahey, T.J., 3rd; Tracey, K.J.; Tekamp-Olson, P.; Cousens, L.S.; Jones, W.G.; Shires, G.T.; Cerami, A.; Sherry, B. Macrophage inflammatory protein 1 modulates macrophage function. *J. Immunol.* **1992**, *148*, 2764–2769. [PubMed]

28. Speyer, C.L.; Gao, H.; Rancilio, N.J.; Neff, T.A.; Huffnagle, G.B.; Sarma, J.V.; Ward, P.A. Novel chemokine responsiveness and mobilization of neutrophils during sepsis. *Am. J. Pathol.* **2004**, *165*, 2187–2196. [CrossRef]

29. Noguchi, Y.; Shinozaki, Y.; Fujishita, K.; Shibata, K.; Imura, Y.; Morizawa, Y.; Gachet, C.; Koizumi, S. Astrocytes protect neurons against methylmercury via ATP/P2Y(1) receptor-mediated pathways in astrocytes. *PLoS ONE* **2013**, *8*, e57898. [CrossRef] [PubMed]

30. Takahashi, T.; Iwai-Shimada, M.; Syakushi, Y.; Kim, M.S.; Hwang, G.W.; Miura, N.; Naganuma, A. Methylmercury induces expression of interleukin-1β and interleukin-19 in mice brains. *Fundam. Toxicol. Sci.* **2015**, *2*, 239–243. [CrossRef]

Article

Effect of Metallothionein-III on Mercury-Induced Chemokine Gene Expression

Jin-Yong Lee [1], Maki Tokumoto [1], Gi-Wook Hwang [2], Min-Seok Kim [2,3], Tsutomu Takahashi [2,4], Akira Naganuma [2], Minoru Yoshida [5] and Masahiko Satoh [1,*]

[1] Laboratory of Pharmaceutical Health Sciences, School of Pharmacy, Aichi Gakuin University, 1-100 Kusumoto-cho, Chikusa-ku, Nagoya 464-8650, Japan; leejy@dpc.agu.ac.jp (J.-Y.L.); maki@dpc.agu.ac.jp (M.T.)

[2] Laboratory of Molecular and Biochemical Toxicology, Graduate School of Pharmaceutical Sciences, Tohoku University, Sendai 980-8578, Japan; gwhwang@m.tohoku.ac.jp (G.-W.H.); assams7@naver.com (M.-S.K.); tsutomu@toyaku.ac.jp (T.T.); naganuma@tohoku.ac.jp (A.N.)

[3] Laboratory Animal Center, Daegu-Gyeongbuk Medical Innovation Foundation, Daegu 41061, Korea

[4] Department of Environmental Health, School of Pharmacy, Tokyo University of Pharmacy and Life Sciences, 1432-1, Horinouchi, Hachioji, Tokyo 192-0392, Japan

[5] Faculty of Health and Medical Care, Hachinohe Gakuin University, 3-98 Mihono, Hachinohe 031-8588, Japan; m2yosida@hachinohe-u.ac.jp

* Correspondence: masahiko@dpc.agu.ac.jp; Tel.: +81-52-757-6790; Fax: +81-52-757-6799

Received: 15 June 2018; Accepted: 7 August 2018; Published: 12 August 2018

Abstract: Mercury compounds are known to cause central nervous system disorders; however the detailed molecular mechanisms of their actions remain unclear. Methylmercury increases the expression of several chemokine genes, specifically in the brain, while metallothionein-III (MT-III) has a protective role against various brain diseases. In this study, we investigated the involvement of MT-III in chemokine gene expression changes in response to methylmercury and mercury vapor in the cerebrum and cerebellum of wild-type mice and MT-III null mice. No difference in mercury concentration was observed between the wild-type mice and MT-III null mice in any brain tissue examined. The expression of *Ccl3* in the cerebrum and of *Cxcl10* in the cerebellum was increased by methylmercury in the MT-III null but not the wild-type mice. The expression of *Ccl7* in the cerebellum was increased by mercury vapor in the MT-III null mice but not the wild-type mice. However, the expression of *Ccl12* and *Cxcl12* was increased in the cerebrum by methylmercury only in the wild-type mice and the expression of *Ccl3* in the cerebellum was increased by mercury vapor only in the wild-type mice. These results indicate that MT-III does not affect mercury accumulation in the brain, but that it affects the expression of some chemokine genes in response to mercury compounds.

Keywords: methylmercury; mercury vapor; metallothionein-III; chemokine

---

1. Introduction

Several mercury compounds are considered hazardous, exerting mainly central nervous system (CNS) or renal damage [1–3]. Methylmercury is a pollutant that causes severe damage [4,5], and accumulates in fish by bioaccumulation. Recent epidemiological investigations have shown that pregnant women who take up relatively large amounts of methylmercury are at higher risk of delivering children with developmental disorders [6,7]. Mercury vapor also produces neuronal damage, and the exposure of gold miners to high levels of mercury vapor through mercury amalgamation is a great concern in developing countries [8,9]. Furthermore, the exposure of children to mercury vapor has increased due to the use of mercury in amalgam to extract gold from ores [10]. Recently, it was reported that high concentrations of methylmercury were detected in grains yielded near a mercury mine in China [11]. These reports suggest that rice is a source of methylmercury exposure in the area. In addition, the median estimated methylmercury intake for children was 0.29 μg/kg bodyweight/week, which is approximately 16% above the dietary references dose (RfD) in Hong Kong [12]. Furthermore, a recent Japanese cohort study suggested that prenatal methylmercury exposure affected neuronal function during child development [13,14].

Metallothionein (MT) is a defense factor against harmful metals such as mercury and cadmium. MT is a cysteine rich, low molecular protein that reduces toxicity by binding to metals [15,16]. There are four isoforms of MT, I to IV, which have different locations and levels of expression and varied physiological functions [15,16]. MT-III, which is present in the brain, has a protective effect against various brain diseases [17–19].

Recently, it has been reported that methylmercury increases the expression of several chemokine genes specific to the mouse brain [20]. Furthermore, we examined the impact of methylmercury on the expression of chemokine genes in various mouse tissues and found that expression of *Ccl4* shows brain specific induction by methylmercury treatment [21]. Chemokines are a type of cytokine known to cause migration of leukocytes. They are secreted primarily from immune cells and participate in the inflammatory response [22]. Chemokines are also secreted from various tissues, including the brain, kidney, and liver [23–26].

In this study, we investigated the involvement of MT-III in chemokine gene expression in the cerebrum and cerebellum in response to methylmercury and mercury vapor using the MT-III null mice.

2. Materials and Methods

2.1. Animals and Exposure Procedures

MT-III null mice and 129/Sv mice as wild-type controls were purchased from The Jackson Laboratory (Bar Harbor, ME, USA) and routinely bred in the vivarium of the School of Pharmacy, Aichi Gakuin University. MT-III null mice were engineered by Erickson et al. [18] and had the 129/Sv genetic background. Three-week-old female mice were caged in a ventilated animal room at 24 °C ± 1 °C with 50 ± 10% relative humidity, and a 12 h light–dark cycle in the animal room of the School of Pharmacy, Aichi Gakuin University. Mice were maintained on standard laboratory food (MF, Oriental Yeast Co., Tokyo, Japan) and tap water ad libitum, and they received humane care throughout the experiment according to the guidelines of the School of Pharmacy.

Mice were assigned randomly to control or experimental groups (n = 4–5). For mercury vapor (Hg^0) exposure, mice were placed in a mercury vapor exposure chamber and exposed for 8 h every day at a mean concentration of 0.121 (range: 0.080 to 0.180) mg/m^3 for 4 weeks. The concentration of mercury in the exposure chamber was measured every day using a mercury survey meter (EMP-1A, Nippon Instruments Co., Tokyo, Japan). For methylmercury (CH_3Hg^+) exposure, methylmercury chloride (GL Sciences Inc., Tokyo, Japan) was diluted with distilled water to prepare a 5 ppm solution. The solution containing 5 ppm methylmercury was given ad libitum instead of tap water. After 4 weeks of exposure, the cerebrum and cerebellum were removed from each mouse under ether anesthesia.

2.2. Real-Time Reverse Transcription-Polymerase Chain Reaction (RT-PCR)

Total RNA was extracted from brain tissues using TRIzol® Reagent (Ambion, Grand Island, NY, USA) according to the manufacturer's instructions. Total RNA was incubated with a PrimeScript™ RT Reagent Kit (Perfect Real Time) (TaKaRa Bio, Shiga, Japan) to generate cDNA. Real time PCR was performed using SYBR Premix Ex Taq™ II (Perfect Real Time) (TaKaRa Bio) and a Thermal Cycler Dice Real time system (TaKaRa Bio). PCR conditions were: 10 s of hot start at 95 °C followed by 40 cycles of 5 s at 95 °C and 30 s at 60 °C. Gene expression was normalized to *β-actin* mRNA levels. Oligonucleotide sequences of the primers (sense and antisense, respectively) were: 5′-TCTAAGCGTCACCACGACTTCA-3′ and 5′-GTGCACTTGCAGTTCTTGCAG-3′ for the mouse *MT-I* gene; 5′-CCTGCAATGCAAACAACAATGC-3′ and 5′-AGCTGCACTTGTCGGAAGC-3′ for the mouse *MT-II* gene; 5'-AGGGCTGCAAATGCACG-3' and 5'-ACACACAGTCCTTGGCACACTTC-3' for the mouse *MT-III* gene; 5'-ATGAAGGTCTCCACCACTGC-3' and 5'-CCCAGGTCTCTTTGG AGTCA-3' for the mouse *Ccl3* gene; 5'-CAAACCTAACCCCGAGCAACAC-3' and 5'-GGTCTCATA GTAATCCATCACAAAGC-3' for the mouse *Ccl4* gene; 5'-AATGCATCCACATGCTGCTA-3' and 5'-CTTTGGAGTTGGGGTTTTCA-3' for the mouse *Ccl7* gene; 5'-GTCCTCAAGGTATTGGCTGGA-3' and 5'-GGGTCAGCACAGATCTCCTT-3' for the mouse *Ccl12* gene; 5'-AAGTGCTGCCGTCATTTTCT-3' and 5'-GTGGCAATGATCTCAACACG-3' for the mouse *Cxcl10* gene; 5'-CCTAAGGCCAACCGTGAAAA-3' and 5'-AGGCATACAGGGACAGCACA-3' for the mouse *β-actin* gene.

2.3. Analysis of Mercury Concentrations in Tissues

Mercury concentrations in tissues were measured with a cold vapor atomic absorption spectrophotometer (RA-3 Mercury Analyzer; Nippon Instruments, Tokyo, Japan) after digestion with a concentrated acid mixture [HNO_3/$HClO_4$ 1:3 (v/v)].

2.4. Statistical Analyses

Statistical analyses were undertaken using single factor ANOVA followed by Bonferroni's test for post hoc comparison ($P < 0.05$).

3. Results

3.1. Body Weight Changes

Body weights of the MT-III null mice and wild-type mice were measured one day after completion of exposure. The body weights of wild-type mice and MT-III null mice exposed to methylmercury were similar to those of the corresponding control mice (Figure 1). The body weights of wild-type mice and MT-III null mice exposed to mercury vapor were significantly lower than those of the corresponding control mice (Figure 1). However, there was no difference in body weight fluctuation in response to mercury exposure between wild-type mice and MT-III null mice (Figure 1).

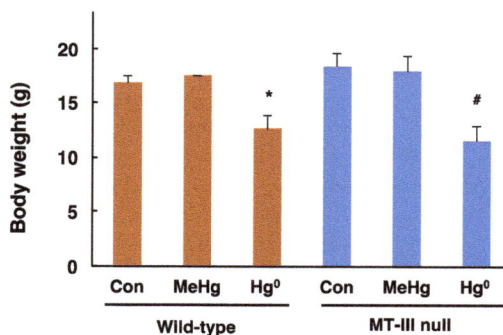

Figure 1. Body weights of the wild-type and MT-III null mice exposed to mercury compounds. Body weights of the wild-type and MT-III null mice were measured one day after the final mercury exposure. Values are the mean ± S.D. (n = 4–5). * Significantly different from the control group of wild-type mice, $P < 0.05$. # Significantly different from the control group of MT-III null mice, $P < 0.05$.

3.2. Total Mercury Concentrations in the Cerebrum and Cerebellum

The total mercury levels in the cerebrum and cerebellum of mice exposed to methylmercury or mercury vapor are shown in Figure 2. In the cerebrum of wild-type mice, mercury concentrations of 2380.57 ± 300.24 ng/g tissue and 519.99 ± 13.68 ng/g tissue were detected after methylmercury exposure and mercury vapor exposure, respectively (Figure 2A). In the cerebrum of MT-III null mice, mercury concentrations of 2373.14 ± 957.12 ng/g tissue and 475.79 ± 56.73 ng/g tissue were detected after methylmercury exposure and mercury vapor exposure, respectively (Figure 2A). In the cerebellum of wild-type mice, mercury concentrations of 1715.23 ± 121.09 ng/g tissue and 783.58 ± 44.15 ng/g tissue were detected following methylmercury exposure and mercury vapor exposure, respectively (Figure 2B). In the cerebellum of MT-III null mice, mercury concentrations of 1643.18 ± 130.82 ng/g tissue and 779.71 ± 54.85 ng/g tissue were detected in response to methylmercury exposure and mercury vapor exposure, respectively (Figure 2B). However, no significant difference was observed between wild-type mice and MT-III null mice in either the cerebrum or cerebellum (Figure 2).

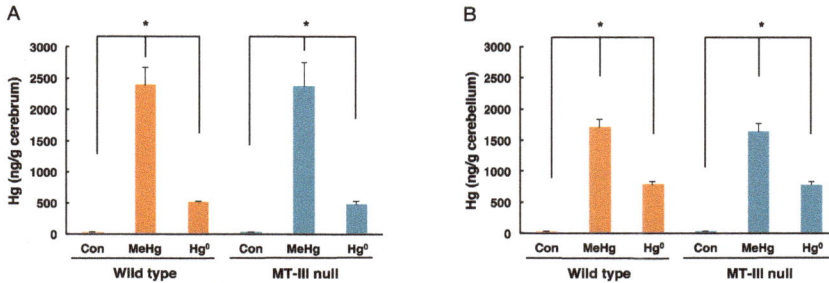

Figure 2. Mercury concentrations in the cerebrum and cerebellum of wild-type mice and MT-III null mice exposed to mercury compounds. Cerebra and cerebella were removed one day after the final mercury exposure. Total mercury levels in the cerebrum (**A**) and cerebellum (**B**) were measured. Values are the mean ± S.D. (n = 4–5). * $P < 0.05$.

3.3. The Levels of MT-I, MT-II, MT-III mRNAs in the Cerebrum and Cerebellum after Mercury Exposure

The mRNA levels of *MT-I* in the cerebrum were significantly increased only in the wild-type mice exposed to methylmercury (Figure 3A). The mRNA levels of *MT-II* in the cerebrum were significantly decreased only in MT-III null mice exposed to mercury vapor (Figure 3B). The *MT-III* mRNA levels in the cerebrum of wild-type mice were not altered by methylmercury or mercury vapor exposure (Figure 3C).

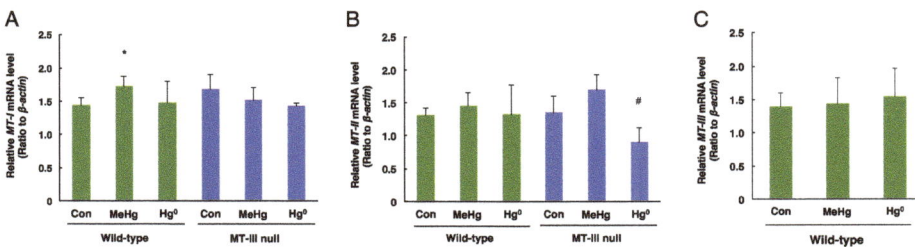

Figure 3. The levels of *MT-I*, *MT-II* and *MT-III* mRNAs in the cerebra of wild-type mice and MT-III null mice exposed to mercury compounds. mRNA levels of *MT-I* (**A**), *MT-II* (**B**) and *MT-III* (**C**) were determined by real time RT-PCR. mRNA levels were normalized with *β-actin*. Values are the mean ± S.D. (n = 4–5). * Significantly different from the control group of wild-type mice, $P < 0.05$. # Significantly different from the control group of MT-III null mice, $P < 0.05$.

The mRNA levels of *MT-I*, *MT-II* and *MT-III* in the cerebellum of each exposed group are shown in Figure 4. *MT-I* mRNA levels in the cerebellum were not changed by mercury exposure in either wild-type mice or MT-III null mice (Figure 4A). The mRNA levels of *MT-II* in the cerebellum were significantly decreased only in the wild-type mice exposed to methylmercury (Figure 4B). No changes in *MT-III* mRNA levels were observed in the wild-type mice exposed to methylmercury or mercury vapor (Figure 4C). These results indicate that MT-III does not affect the accumulation of mercury after methylmercury or mercury vapor exposure.

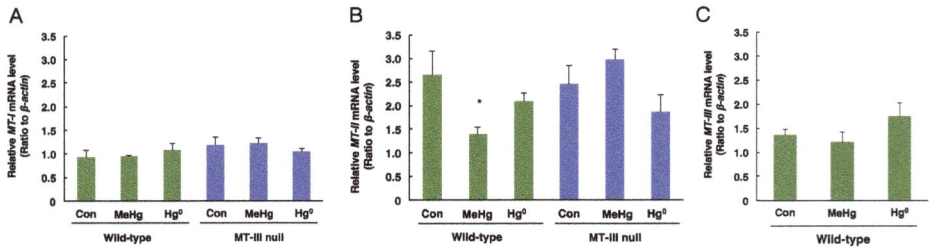

Figure 4. The levels of *MT-I*, *MT-II* and *MT-III* mRNAs in the cerebella of wild-type mice and MT-III null mice exposed to mercury compounds. mRNA levels of *MT-I* (**A**), *MT-II* (**B**) and *MT-III* (**C**) were determined by real time RT-PCR. mRNA levels were normalized with *β-actin*. Values are the mean ± S.D. (n = 4–5). * Significantly different from the control group of wild-type mice, $P < 0.05$.

3.4. Changes in Expression of Chemokine Genes in the Cerebrum in Response to Mercury Exposure

Ccl3 mRNA levels in the cerebrum were significantly increased by methylmercury exposure only in the MT-III null mice (Figure 5A). Moreover, *Ccl3* mRNA levels in the cerebrum of MT-III null mice were higher than those of wild-type mice by methylmercury exposure (Figure 5A). *Ccl4* mRNA levels did not change after mercury exposure (Figure 5B). *Ccl7* mRNA levels were markedly elevated in the MT-III null mice and wild-type mice after methylmercury and mercury vapor exposure (Figure 5C). Although *Ccl12* mRNA levels were increased by MT-III deficiency, methylmercury exposure increased *Ccl12* mRNA levels only in the wild-type mice (Figure 5D). *Cxcl10* mRNA levels were significantly elevated only when the wild-type mice were exposed to methylmercury (Figure 5E).

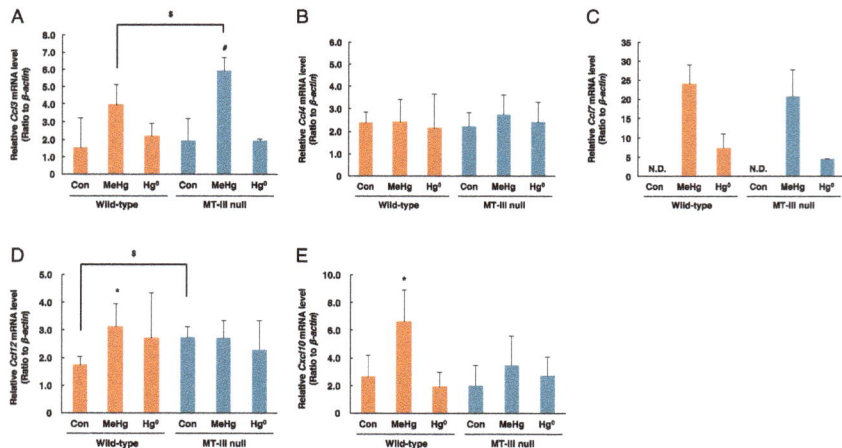

Figure 5. mRNA levels of chemokine genes in the cerebra of wild-type mice and MT-III null mice exposed to mercury compounds. mRNA levels of *Ccl3* (**A**), *Ccl4* (**B**), *Ccl7* (**C**), *Ccl12* (**D**) and *Cxcl10* (**E**) were determined by real-time RT-PCR. mRNA levels were normalized with *β-actin*. Values are the mean ± S.D. (n = 4–5). * Significantly different from the control group of wild-type mice, $P < 0.05$. # Significantly different from the control group of MT-III null mice, $P < 0.05$. $ Significantly different between wild-type and MT-III null mice, $P < 0.05$.

3.5. Changes in Expression of Chemokine Genes in the Cerebellum after Mercury Exposure

The mRNA levels of *Ccl3* in the cerebellum were significantly increased by methylmercury in the MT-III null mice and wild-type mice, and mercury vapor significantly increased *Ccl3* mRNA levels only in the wild-type mice (Figure 6A). *Ccl4* mRNA levels were significantly increased by methylmercury in the MT-III null mice and wild-type mice (Figure 6B). *Ccl7* mRNA levels were significantly increased by methylmercury in the MT-III null mice and wild-type mice, and were significantly increased only in the MT-III null mice in response to mercury vapor (Figure 6C). *Ccl12* mRNA levels were significantly increased by both methylmercury and mercury vapor in the MT-III null mice and wild-type mice (Figure 6D). However, *Ccl12* mRNA levels in the MT-III null mice exposed to methylmercury were lower than those of wild-type mice (Figure 6D). *Cxcl10* mRNA levels were significantly elevated only when the MT-III null mice were exposed to methylmercury (Figure 6E).

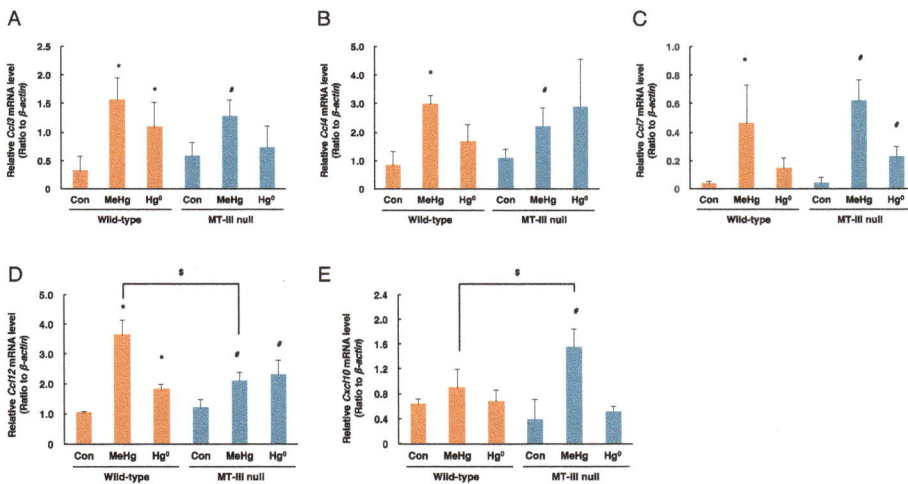

Figure 6. mRNA levels of chemokine genes in the cerebella of wild-type mice and MT-III null mice exposed to mercury compounds. mRNA levels of *Ccl3* (**A**), *Ccl4* (**B**), *Ccl7* (**C**), *Ccl12* (**D**) and *Cxcl10* (**E**) were determined by real-time RT-PCR. mRNA levels were normalized with *β-actin*. Values are the mean ± S.D. (n = 4–5). * Significantly different from the control group of wild-type mice, $P < 0.05$. # Significantly different from the control group of MT-III null mice, $P < 0.05$. $ Significantly different between wild-type and MT-III null mice, $P < 0.05$.

4. Discussion

MT-III is highly expressed in neurons [17], and also shows a protective effect against brain diseases [18,19]. Mercury compounds, such as methylmercury and mercury vapor, are harmful substances that cause disorders of the CNS, but the influence of MT-III on the disorders caused by these mercuric compounds is poorly understood. Recently, methylmercury has been reported to increase expression of the chemokine genes, *Ccl3*, *Ccl4*, *Ccl7*, *Ccl12* and *Cxcl10* in the mouse brain [20,21,27]. The most highlighted knowledge in this study is that *Ccl3* expression in the cerebrum and the cerebellar expression of *Ccl12* and *Cxcl10* were significantly different between wild-type mice and MT-III null mice only in the methylmercury exposure. MT-III is abundantly expressed in the normal brain; on the other hand, in a brain with Alzheimer's disease, MT-III expression is largely reduced [17]. It was suggested that MT-III might have a protective role in cerebral ischemia [28]. Furthermore, inflammatory factors are upregulated upon the brain damage, as it increases phagocytosis and the release of inflammatory mediators [29,30]. The present study suggests that the increase in chemokine expression may be involved in methylmercury-triggered brain damage with topical specificity. The present study also suggest that MT-III may be the cause of the difference in inflammatory response upon methylmercury exposure.

Our present study revealed that the MT-III deficiency had no effect on mercury concentration in the brains of mice exposed to mercury compounds; however, the MT-III deficiency influenced the expression of chemokine genes in response to mercury exposure. The expression of *Ccl12* and *Cxcl10* in the cerebrum was increased in the wild-type mice by methylmercury; but not in the cerebrum of MT-III null mice. Moreover, in response to mercury vapor exposure, *Ccl3* expression in the cerebellum was not changed in the MT-III null mice, but was significantly elevated in the wild-type mice. In the cerebellum of wild-type mice, the expression of *Ccl7* was not changed by mercury vapor exposure and that of *Cxcl10* was not changed by methylmercury exposure; however, they were significantly elevated in the MT-III null mice. Indeed, the fluctuation in *Cxcl10* expression in response to methylmercury was observed in the cerebrum of wild-type mice, and in the cerebellum of MT-III null mice. Taken together, these findings indicate that MT-III may affect the expression of some chemokine genes in response to different mercury compounds differently in the cerebrum and cerebellum. In recent years, MT-I and MT-II have been reported to be involved in inflammatory reactions in cardiomyocytes [31]. The present study indicates that MT-III may be involved in inflammatory reactions consequent to an exogenous stress on the brain.

Chemokines are associated with inflammatory damage in a number of tissues, including the brain [23–26]. Chemokines may function as signaling molecules in the CNS [32]. Accordingly, the expression of several chemokines increases with hypoglossal nerve damage, however the role of chemokines in cranial nerves remains unclear. Methylmercury induces a brain-specific increase in *Ccl4* expression in mice [21]. In addition, it is assumed that CCL4 is involved in the pathway mediating the toxicity of methylmercury in the CNS [33,34]. In this study, methylmercury induced *Ccl4* gene expression in the cerebellum not only in the wild-type mice but also in the MT-III null mice. Therefore, CCL4 may be an important factor mediating methylmercury toxicity.

Previous studies revealed that postnatal mice exposed to 0.188 mg/m^3 mercury vapor or 3.85 ppm methylmercury showed a decrease in total locomotive activity in the OPF [35]. In this study, we exposed the mice to similar doses of mercury compounds. Because several chemokine gene expressions were increased in brain tissues of wild-type mice, chemokine genes may be associated with the influence of mercury exposure in childhood on neuronal development.

MT-I/II null mice showed an enhanced adverse effect of neonatal mercury (HgCl$_2$) exposure [19]. Moreover, MT-I/II null mice exposed to mercury vapor showed decreased locomotor activity [36,37]. Furthermore, the highly sensitive methylmercury toxicity in MT-I/II deficient astrocytes was rescued by the introduction of the MT-I gene [38]. Thus, MT-I and MT-II were suggested to be involved in mercury compound induced neuropathy, but the role of MT-III in neuropathy caused by mercury compounds is almost unknown. Our present study revealed that MT-III deficiency is involved in mercury-induced neuropathy through changing cerebral chemokine gene expression.

Author Contributions: J.-Y.L. conducted experiments, analyzed the data and wrote the manuscript. M.T. analyzed the gene expression of metallothioneins. G.-W.H., M.-S.K., T.T., and A.N. analyzed gene expression of chemokines. M.Y. supported mercury exposure and mercury analyses. M.S. managed the project and revised the manuscript.

Funding: This research was supported by a research grant from the Institute of Pharmaceutical Life Sciences, Aichi Gakuin University.

Acknowledgments: We sincerely thank Ryutaro Kimata for excellent experimental support.

Conflicts of Interest: The authors declare no conflicts of interest.

References

1. Lee, J.Y.; Ishida, Y.; Kuge, S.; Naganuma, A.; Hwang, G.W. Identification of substrates of F-box protein involved in methylmercury toxicity in yeast cells. *FEBS Lett.* **2015**, *589*, 2720–2725. [CrossRef] [PubMed]

2. Lee, J.Y.; Ishida, Y.; Takahashi, T.; Naganuma, A.; Hwang, G.W. Transport of pyruvate into mitochondria is involved in methylmercury toxicity. *Sci. Rep.* **2016**, *6*, 21528. [CrossRef] [PubMed]

3. Tokumoto, M.; Lee, J.Y.; Shimada, A.; Tohyama, C.; Satoh, M. Glutathione has a more important role than metallothionein-I/II against inorganic mercury-induced acute renal toxicity. *J. Toxicol. Sci.* **2018**, *43*, 275–280. [CrossRef] [PubMed]

4. Mori, N.; Yasutake, A.; Marumoto, M.; Hirayama, K. Methylmercury inhibits electron transport chain activity and induces cytochrome c release in cerebellum mitochondria. *J. Toxicol. Sci.* **2011**, *36*, 253–259. [CrossRef] [PubMed]

5. Watanabe, J.; Nakamachi, T.; Ogawa, T.; Naganuma, A.; Nakamura, M.; Shioda, S.; Nakajo, S. Characterization of antioxidant protection of cultured neural progenitor cells (NPC) against methylmercury (MeHg) toxicity. *J. Toxicol. Sci.* **2009**, *34*, 315–325. [CrossRef] [PubMed]

6. Grandjean, P.; Weihe, P.; Jørgensen, P.J.; Clarkson, T.; Cernichiari, E.; Viderø, T. Impact of maternal seafood diet on fetal exposure to mercury, selenium, and lead. *Arch. Environ. Health* **1992**, *47*, 185–195. [CrossRef] [PubMed]

7. Grandjean, P.; Weihe, P.; White, R.F.; Debes, F.; Araki, S.; Yokoyama, K.; Murata, K.; Sørensen, N.; Dahl, R.; Jørgensen, P.J. Cognitive deficit in 7-year-old children with prenatal exposure to methylmercury. *Neurotoxicol. Teratol.* **1997**, *19*, 417–428. [CrossRef]

8. Hinton, J.J.; Veiga, M.M.; Veiga, A.T.C. Clean artisanal gold mining: A utopian approach? *J. Clean. Prod.* **2003**, *11*, 99–115. [CrossRef]

9. Veiga, M.M.; Maxson, P.A.; Hylander, L.D. Origin and consumption of mercury in small-scale gold mining. *J. Clean. Prod.* **2006**, *14*, 436–447. [CrossRef]

10. Bose-O'Reilly, S.; Lettmeier, B.; Gothe, R.M.; Beinhoff, C.; Siebert, U.; Drasch, G. Mercury as a serious health hazard for children in gold mining areas. *Environ. Res.* **2008**, *107*, 89–97. [CrossRef] [PubMed]

11. Li, P.; Feng, X.; Qiu, G.; Shang, L.; Wang, S. Mercury exposure in the population from Wuchuan mercury mining area, Guizhou, China. *Sci. Total Environ.* **2008**, *395*, 72–79. [CrossRef] [PubMed]

12. Chan, P.H.Y.; Chan, M.H.M.; Li, A.M.; Cheung, R.C.K.; Yu, X.T.; Lam, H.S. Methylmercury levels in commonly consumed fish and methylmercury exposure of children and women of childbearing age in Hong Kong, a high fish consumption community. *Environ. Res.* **2018**, *166*, 418–426. [CrossRef] [PubMed]

13. Suzuki, K.; Nakai, K.; Sugawara, T.; Nakamura, T.; Ohba, T.; Shimada, M.; Hosokawa, T.; Okamura, K.; Sakai, T.; Kurokawa, N.; et al. Neurobehavioral effects of prenatal exposure to methylmercury and pcbs, and seafood intake: Neonatal behavioral assessment scale results of tohoku study of child development. *Environ. Res.* **2010**, *110*, 699–704. [CrossRef] [PubMed]

14. Tatsuta, N.; Murata, K.; Iwai-Shimada, M.; Yaginuma-Sakurai, K.; Satoh, H.; Nakai, K. Psychomotor ability in children prenatally exposed to methylmercury: The 18-month follow-up of tohoku study of child development. *Tohoku J. Exp. Med.* **2017**, *242*, 1–8. [CrossRef] [PubMed]

15. Klaassen, C.D.; Liu, J.; Choudhuri, S. Metallothionein: An intracellular protein to protect against cadmium toxicity. *Annu. Rev. Pharmacol. Toxicol.* **1999**, *39*, 267–294. [CrossRef] [PubMed]

16. Vašák, M.; Meloni, G. Chemistry and biology of mammalian metallothioneins. *J. Biol. Inorg. Chem.* **2011**, *16*, 1067–1078. [CrossRef] [PubMed]

17. Uchida, Y.; Takio, K.; Titani, K.; Ihara, Y.; Tomonaga, M. The growth inhibitory factor that is deficient in the Alzheimer's disease brain is a 68 amino acid metallothionein-like protein. *Neuron* **1991**, *7*, 337–347. [CrossRef]

18. Erickson, J.C.; Hollopeter, G.; Thomas, S.A.; Froelick, G.J.; Palmiter, R.D. Disruption of the metallothionein-III gene in mice: Analysis of brain zinc, behavior, and neuron vulnerability to metals, aging, and seizures. *J. Neurosci.* **1997**, *17*, 1271–1281. [CrossRef] [PubMed]

19. West, A.K.; Hidalgo, J.; Eddins, D.; Levin, E.D.; Aschner, M. Metallothionein in the central nervous system: Roles in protection, regeneration and cognition. *Neurotoxicology* **2008**, *29*, 489–503. [CrossRef] [PubMed]

20. Hwang, G.W.; Lee, J.Y.; Ryoke, K.; Matsuyama, F.; Kim, J.M.; Takahashi, T.; Naganuma, A. Gene expression profiling using DNA microarray analysis of the cerebellum of mice treated with methylmercury. *J. Toxicol. Sci.* **2011**, *36*, 389–391. [CrossRef] [PubMed]

21. Lee, J.Y.; Hwang, G.W.; Kim, M.S.; Takahashi, T.; Naganuma, A. Methylmercury induces a brain-specific increase in chemokine CCL4 expression in mice. *J. Toxicol. Sci.* **2012**, *37*, 1279–1282. [CrossRef] [PubMed]

22. Zlotnik, A.; Yoshie, O. Chemokines: A new classification system and their role in immunity. *Immunity* **2000**, *12*, 121–127. [CrossRef]
23. Gorter, J.A.; van Vliet, E.A.; Aronica, E.; Breit, T.; Rauwerda, H.; Lopes da Silva, F.H.; Wadman, W.J. Potential new antiepileptogenic targets indicated by microarray analysis in a rat model for temporal lobe epilepsy. *J. Neurosci.* **2006**, *26*, 11083–11110. [CrossRef] [PubMed]
24. Soria, G.; Ben-Baruch, A. The inflammatory chemokines CCL2 and CCL5 in breast cancer. *Cancer Lett.* **2008**, *267*, 271–285. [CrossRef] [PubMed]
25. Saiman, Y.; Friedman, S.L. The role of chemokines in acute liver injury. *Front. Physiol.* **2012**, *3*, 213. [CrossRef] [PubMed]
26. Shimizu, H.; Bolati, D.; Higashiyama, Y.; Nishijima, F.; Shimizu, K.; Niwa, T. Indoxyl sulfate upregulates renal expression of MCP-1 via production of ROS and activation of NF-kappaB, p53, ERK, and JNK in proximal tubular cells. *Life Sci.* **2012**, *90*, 525–530. [CrossRef] [PubMed]
27. Kim, M.S.; Takahashi, T.; Lee, J.Y.; Hwang, G.W.; Naganuma, A. Global chemokine expression in methylmercury-treated mice: Methylmercury induces brain-specific expression of CCL3 and CCL4. *J. Toxicol. Sci.* **2013**, *38*, 925–929. [CrossRef] [PubMed]
28. Koumura, A.; Hamanaka, J.; Shimazawa, M.; Honda, A.; Tsuruma, K.; Uchida, Y.; Hozumi, I.; Satoh, M.; Inuzuka, T.; Hara, H. Metallothionein-III knockout mice aggravates the neuronal damage after transient focal cerebral ischemia. *Brain Res.* **2009**, *1292*, 148–154. [CrossRef] [PubMed]
29. Poulsen, C.B.; Penkowa, M.; Borup, R.; Nielsen, F.C.; Cáceres, M.; Quintana, A.; Molinero, A.; Carrasco, J.; Giralt, M.; Hidalgo, J. Brain response to traumatic brain injury in wild-type and interleukin-6 knockout mice: A microarray analysis. *J. Neurochem.* **2005**, *92*, 417–432. [CrossRef] [PubMed]
30. Scuderi, C.; Stecca, C.; Iacomino, A.; Steardo, L. Role of astrocytes in major neurological disorders: The evidence and implications. *IUBMB Life* **2013**, *65*, 957–961. [CrossRef] [PubMed]
31. Duerr, G.D.; Dewald, D.; Schmitz, E.J.; Verfuerth, L.; Keppel, K.; Peigney, C.; Ghanem, A.; Welz, A.; Dewald, O. Metallothioneins 1 and 2 modulate inflammation and support remodeling in ischemic cardiomyopathy in mice. *Mediat. Inflamm.* **2016**, *2016*, 7174127. [CrossRef] [PubMed]
32. Gamo, K.; Kiryu-Seo, S.; Konishi, H.; Aoki, S.; Matsushima, K.; Wada, K.; Kiyama, H. G-protein-coupled receptor screen reveals a role for chemokine receptor CCR5 in suppressing microglial neurotoxicity. *J. Neurosci.* **2008**, *28*, 11980–11988. [CrossRef] [PubMed]
33. Adler, M.W.; Rogers, T.J. Are chemokines the third major system in the brain? *J. Leukoc. Biol.* **2005**, *78*, 1204–1209. [CrossRef] [PubMed]
34. Miller, R.J.; Rostene, W.; Apartis, E.; Banisadr, G.; Biber, K.; Milligan, E.D.; White, F.A.; Zhang, J. Chemokine action in the nervous system. *J. Neurosci.* **2008**, *28*, 11792–11795. [CrossRef] [PubMed]
35. Yoshida, M.; Lee, J.Y.; Satoh, M.; Watanabe, C. Neurobehavioral effects of postnatal exposure to low-level mercury vapor and/or methylmercury in mice. *J. Toxicol. Sci.* **2018**, *43*, 11–17. [CrossRef] [PubMed]
36. Yoshida, M.; Watanabe, C.; Horie, K.; Satoh, M.; Sawada, M.; Shimada, A. Neurobehavioral changes in metallothionein-null mice prenatally exposed to mercury vapor. *Toxicol. Lett.* **2005**, *155*, 361–368. [CrossRef] [PubMed]
37. Yoshida, M.; Watanabe, C.; Kishimoto, M.; Yasutake, A.; Satoh, M.; Sawada, M.; Akama, Y. Behavioral changes in metallothionein-null mice after the cessation of long-term, low-level exposure to mercury vapor. *Toxicol. Lett.* **2006**, *161*, 210–218. [CrossRef] [PubMed]
38. Yao, C.P.; Allen, J.W.; Mutkus, L.A.; Xu, S.B.; Tan, K.H.; Aschner, M. Foreign metallothionein-I expression by transient transfection in MT-I and MT-II null astrocytes confers increased protection against acute methylmercury cytotoxicity. *Brain Res.* **2000**, *855*, 32–38. [CrossRef]

Article

Sub-Nanomolar Methylmercury Exposure Promotes Premature Differentiation of Murine Embryonic Neural Precursor at the Expense of Their Proliferation

Xiaoyang Yuan [1], Jing Wang [2,3,4] and Hing Man Chan [1,*]

[1]   Department of Biology, University of Ottawa, Ottawa, ON K1N 6N5, Canada; michelle_yuan@hotmail.com

[2]   Regenerative Medicine Program, Ottawa Hospital Research Institute, Ottawa, ON K1H 8L6, Canada; JIWang@ohri.ca

[3]   Department of Cellular and Molecular Medicine, University of Ottawa, Ottawa, ON K1H 8M5, Canada

[4]   Brain and Mind Research Institute, University of Ottawa, Ottawa, ON K1H 8M5, Canada

*   Correspondence: laurie.chan@uottawa.ca; Tel.: +1-613-562-5800 (ext. 7116)

Received: 7 September 2018; Accepted: 3 October 2018; Published: 10 October 2018

Abstract: Methylmercury (MeHg) is a ubiquitous environmental pollutant that is known to be neurotoxic, particularly during fetal development. However, the mechanisms responsible for MeHg-induced changes in adult neuronal function, when their exposure occurred primarily during fetal development, are not yet understood. We hypothesized that fetal MeHg exposure could affect neural precursor development leading to long-term neurotoxic effects. Primary cortical precursor cultures obtained from embryonic day 12 were exposed to 0 nM, 0.25 nM, 0.5 nM, 2.5 nM, and 5 nM MeHg for 48 or 72 h. Total Hg accumulated in the harvested cells in a dose-dependent manner. Not all of the concentrations tested in the study affected cell viability. Intriguingly, we observed that cortical precursor exposed to 0.25 nM MeHg showed increased neuronal differentiation, while its proliferation was inhibited. Reduced neuronal differentiation, however, was observed in the higher dose groups. Our results suggest that sub-nanomolar MeHg exposure may deplete the pool of neural precursors by increasing premature neuronal differentiation, which can lead to long-term neurological effects in adulthood as opposed to the higher MeHg doses that cause more immediate toxicity during infant development.

Keywords: methylmercury; cortical precursors; developmental neurotoxicity; low-dose exposure; delayed effects

1. Introduction

Methylmercury (MeHg) is a global pollutant affecting millions of people worldwide [1,2]. Its main target is the central nervous system, with fetuses being particularly susceptible [3]. Amin-Zaki et al. [4] reported that the levels of MeHg in fetal blood are about 25% higher than those of the mother. It has also been shown that fetuses can be affected in the absence of maternal toxicity [5].

A cohort study on the population of the Faroe Islands showed that prenatal exposure to MeHg was significantly associated with deficits in fine motor control, language, and learning abilities in children and adolescents [6]. Yorifuji et al. [7] demonstrated an increased prevalence of psychiatric symptoms in adults who showed no sign of toxicity at birth, based on an epidemiological study on the residents of Minamata.

Developmental exposure to MeHg in mice results in memory disturbances and induces depression-like behavior in adult animals [8], which persists into older age [9]. Animal studies have shown that the developing brain is extremely vulnerable to MeHg neurotoxicity [8–10], which may be attributed to the rapid cell proliferation [11] and cell differentiation in the developing brain [12,13].

Previous studies have demonstrated detrimental effects of MeHg at micromolar levels on neurogenesis, including inhibition of proliferation and disturbed cell cycle progression in neuronal cells [14].

The emerging idea of the "Developmental Origin of Health and Diseases (DOHaD)", previously termed the "fetal origins of adult disease" in the 1990s, postulates that exposure to environmental influences during the embryonic period is related to the risk of developing diseases in adulthood [15]. Therefore, it is important to understand not only the immediate effect of MeHg exposure on the embryos themselves, but also its potential influences that progress into adulthood. The DOHaD theory suggests that epigenetic alterations could be induced by environmental conditions during development, which are maintained in adulthood. These subtle epigenetic changes, showing no effect in early ages, can increase the risk of developing diseases later in life [16]. Therefore, there is the necessity of understanding the effect of dietary MeHg exposure on very low levels of embryonic neural development and its underlying mechanisms.

In this study, we investigated the effects of low-dose (nM) MeHg exposure on regulating murine embryonic neural precursor development using a mouse cerebral cortex development model. Our hypothesis was that a low and non-toxic dose of MeHg could disrupt the development of these cortical precursors leading to long-term neurotoxic effects.

2. Materials and Methods

2.1. Animal Ethics

Animal care, handling and use protocols were reviewed and approved by the Animal Care Committee of the Ottawa Hospital Research Institute, University of Ottawa on 28 July, 2017. Animal protocol number: OHRI-2103.

2.2. Cell Culture Procedures and Experimental Treatments

Primary cultures of cortical precursors were obtained from embryonic day 12 cortices dissected in ice-cold Hanks' balanced salt solution (HBSS) (Life Technologies, Carlsbad, CA, USA) from CD1 mouse (Charles River Laboratories, Wilmington, MA, USA). Embryos were transferred to ice-cold HBSS and the cerebral cortices were carefully isolated from the brain after removing the meninges. The tissue was mechanically triturated with a plastic pipette and plated at a density of 10^5 cells on coverslips pre-coated with 15% poly-L-ornithine (PLO) (Sigma, St. Louis, MO, USA) and 5% laminin (BD Biosciences, Franklin Lakes, NJ, USA) in a 24-well plate (Thermo Scientific BioLite, Waltham, MA, USA). The cortical precursors were cultured in a neurobasal medium (Invitrogen, Carlsbad, CA, USA) containing 500 μM L-glutamine (Cambrex Biosciences, East Rutherford, NJ, USA), 2% B27 supplement (Invitrogen, Carlsbad, CA, USA), 1% penicillin-streptomycin (Invitrogen, Carlsbad, CA, USA), and 40 ng/mL FGF2 (BD Biosciences, Franklin Lakes, NJ, USA). The primary culture was exposed to 0 nM, 0.25 nM, 0.5 nM, 2.5 nM, and 5 nM MeHg for 48 or 72 h. The dosing solution was freshly prepared daily from a stock solution of 1 M using MeHgCl from Alfa Aesar (Ward Hill, MA, USA).

2.3. Immunocytochemistry

Cultured cells were fixed in 4% paraformaldehyde for 10 min, then blocked with 10% normal goat serum (NGS) diluted in PBS with 0.3% Triton X-100. Primary antibodies were diluted in 10% NGS in PBS with 0.3% Triton X-100 and incubated in a humid chamber at 4 °C overnight. Secondary antibodies were diluted in PBST and incubated for 1 h at room temperature. Hoechst 33258 was diluted in PBS. The culture was washed three times for 5 min with PBS between each step. The primary antibodies used were mouse anti-βIII-tubulin (1:1000; Covance, Princeton, NJ, USA), rabbit anti-Pax-6 (1:2000; BioLegend, San Diego, CA, USA), rabbit anti-Sox2 (1:100; Millipore, Burlington, MA, USA), mouse anti-Ki67 (1:400; Abcam, Cambridge, UK), and rabbit anti-Cleaved Caspase-3 (CC3) (1:400; Cell Signalling Tech., Danvers, MA, USA). The secondary antibodies used were Alexa Fluor 555- and

Alexa Fluor 488-conjugated goat antibodies (1:500; Life Technologies, Carlsbad, CA, USA). Nuclear staining was performed with Hoechst 33343 (1:1000; Sigma, St. Louis, MO, USA). After rinsing with PBS, the coverslips were mounted in a Lab Vision PermaFluor Aqueous Mounting Medium (Thermo Fisher, Waltham, MA, USA). All experiments were repeated at least three times.

2.4. Microscopy and Quantification

Digital image acquisition was performed using a Zeiss Axioplan 2 fluorescent microscope with Zeiss Axiovision software (Carl Zeiss Microscopy, Thornwood, NY, USA). Six random images over 300 cells per condition per experiment were taken for quantification.

2.5. Statistics

All data were expressed as the mean plus or minus the standard error of the mean (SEM) and were tested for statistical significance with one-way ANOVA, followed by Bonferroni's post hoc test. A two-tailed Student's t-test was used for between-group comparisons. The differences were considered significant if $p < 0.05$. All statistical analyses were performed using Prism (version 7, GraphPad Software, La Jolla, CA, USA, 2018).

3. Results

3.1. Cell Viability

Immunocytochemistry results showed that the percentage of CC3+ condensed nuclei was not changed between the methylmercury treated groups and the control group (Figure 1A). One-way ANOVA results showed that MeHg treatments did not have any effect on CC3+, F = 0.4666, d.f. = (4,10), $p > 0.05$) (Figure 1B). Thus, the exposure levels of methylmercury used in our experiments did not the affect cell viability of the cortical precursors.

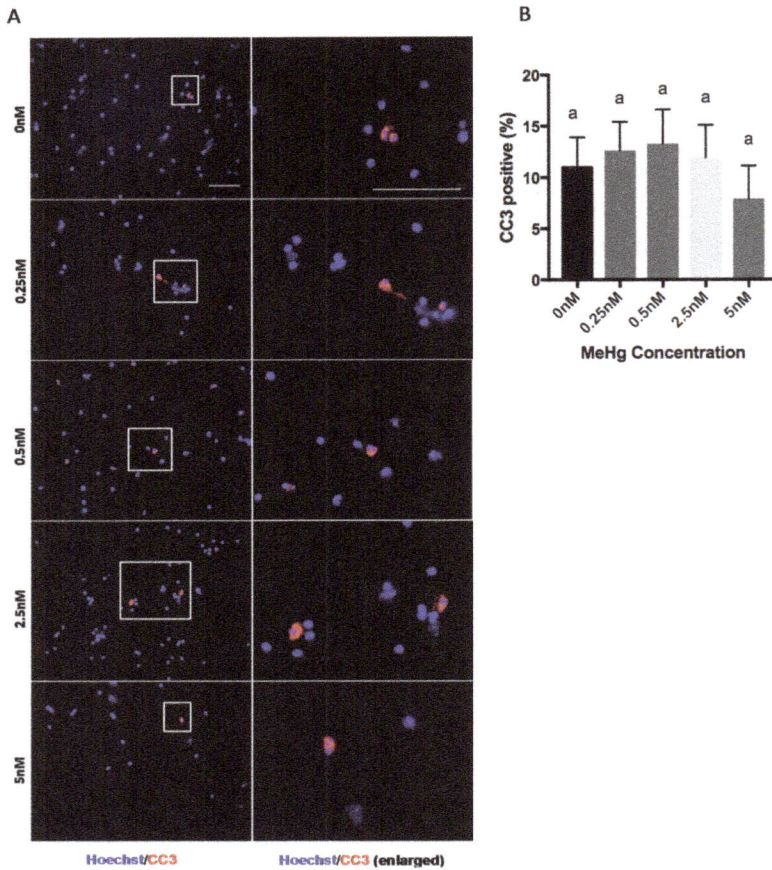

Figure 1. Methylmercury treatments at doses of 0.25 to 5 nM do not induce cell death. Immunocytochemistry was performed in a 2-day cortical precursors culture. Images of Cleaved Caspase-3 (CC3) (Red) and Hoechst (Blue) staining are shown. Scale bar = 50 μm (**A**). The bar graph (**B**) shows the percentage of immunocytochemistry-positive cells. Values are mean ± SEM (*n* = 3). Statistical significance was determined by a one-way ANOVA followed by a Bonferroni's post hoc test. No significant *p* value was obtained for ANOVA. Letters denote the results of the comparisons between treatment groups, groups with the same letter were not statistically different.

3.2. Effects of Cortical Precursors on Proliferation and Differentiation

The MeHg treatments had a significant effect on the proliferation and differentiation of cortical precursors. One-way ANOVA results showed that there was a significant MeHg treatment effect for Pax6 (F = 13.56, d.f. = (4,10), $p < 0.05$) and Beta III tubulin staining (F = 50.12, d.f. = (4,10), $p < 0.05$). Immunofluorescence results showed that exposure to 0.25 nM MeHg significantly increased the percentage of new born neurons produced from E12 cortical precursors, labeled with βIII tubulin, compared to the control (Figure 2A). Coincidentally, the population of Pax6 + cortical precursors was significantly decreased at 0.25 nM MeHg (Figure 2B). To validate the reduced pool of cortical precursors in the culture, we performed immunocytochemistry analysis with a pan-neural stem cell marker, Sox2, and a cell cycling marker, Ki67. One-way ANOVA results showed that there was a significant MeHg treatment effect for Sox2 (F = 8.849, d.f. = (4,10), $p < 0.05$) and Ki67 staining (F = 5.182, d.f. = (4,10),

$p < 0.05$). The results showed that both the percentage of Sox2+ cortical precursors and Ki67+ cycling cells were dramatically decreased upon exposure to 0.25 nM MeHg (Figure 2D–F). These results suggest that exposure of cortical precursors to an extremely low dose (0.25 nM) of MeHg enhances premature neuronal differentiation while reducing their proliferation. In comparison, exposure of cortical precursors to MeHg from 0.5 nM to 5 nM reduced the percentage of βIII tubulin+ new born neurons in culture (Figure 2A,C). However, while the 0.5 nM and 2.5 nM treatment groups showed a significant decrease in the population of Pax6+ cortical precursors, the 5 nM treatment group did not (Figure 2B). In addition, the percentage of Sox2+ NSCs and Ki67+ cycling cells were not changed in the 0.5 nM and 5 nM MeHg treatment groups (Figure 2D–F). These results show that exposure to 0.5 nM did not show the effects that we observed at the lower dose of 0.25 nM. Its effect was more similar to the effects observed at 2.5 and 5 nM MeHg, that showed significantly lower neuronal differentiation, while its proliferation recovered gradually to a level comparable with the control.

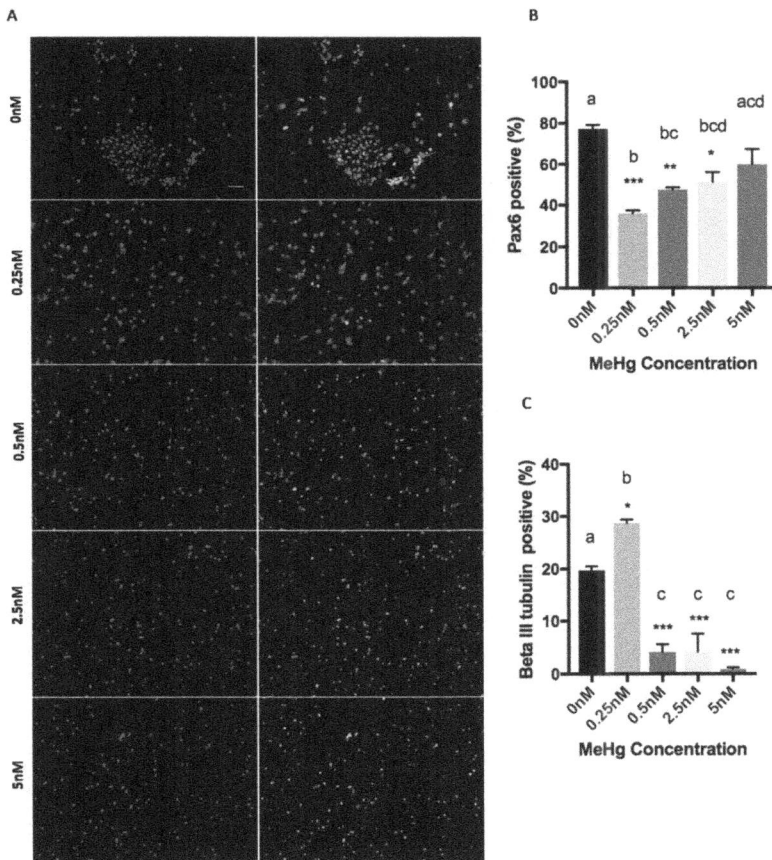

Figure 2. *Cont.*

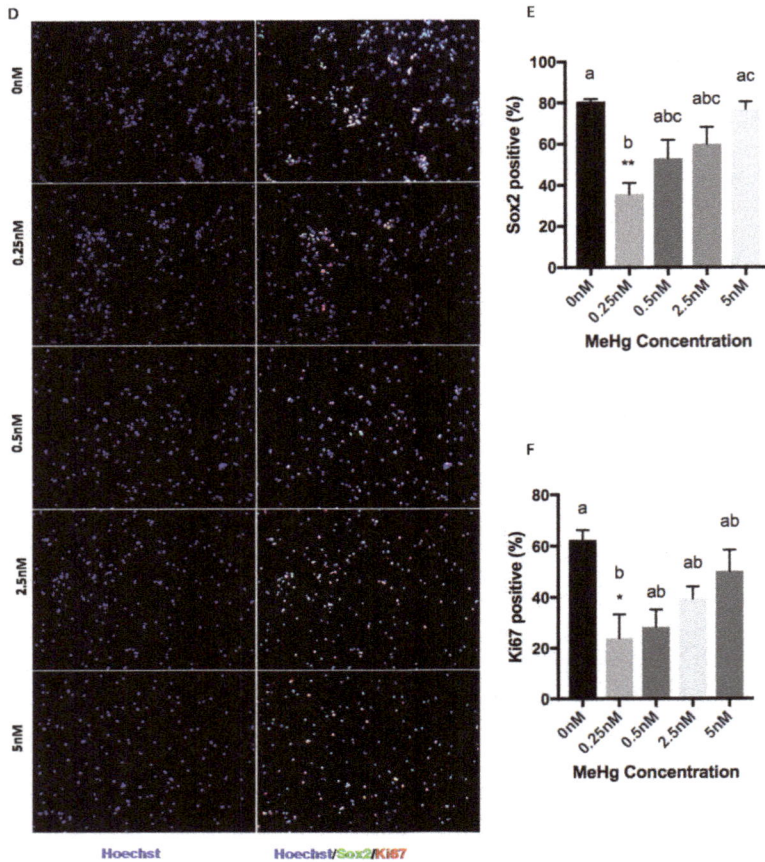

Figure 2. Immunocytochemistry was performed in a 3-day NSC culture. Images of (**A**) Pax6 (Green), Beta III Tubulin (Red), (**D**) Sox2 (Green), Ki67 (Red), and Hoechst (Blue) staining are shown. Scale bar = 50 µm. The bar graphs (**B,C**; **E,F**) show the percentage of immunocytochemistry-positive cells. Values are mean ± SEM (*n* = 3). Statistical significance was determined by a one-way ANOVA followed by a Bonferroni's post hoc test (* *p* < 0.05: vs. control, letters denote the results of the comparisons between treatment groups, groups with the same letter were not statistically different).

4. Discussion

The novel finding of this study is that the effects of MeHg on cortical precursor development are dose-dependent. The extremely low sub-nanomolar dose of 0.25 nM MeHg increases the neuronal differentiation of cortical precursors while reducing their proliferation. On the other hand, there was a decrease in differentiation at higher doses (>0.5 nM), which is similar to the results reported in the existing literature. Similar doses of MeHg (2.5–5 nM MeHg for 48 h) were shown to inhibit the spontaneous neuronal differentiation of murine embryonic neural stem cells [12]. Fujimura & Usuki [17] also showed that neural progenitor cell proliferation was suppressed 48 h after exposure to 10 nM MeHg, but cell death was not observed. Tamm et al. [18] identified Notch signaling as a target for methylmercury's inhibition of neuronal differentiation at exposure levels between 2.5 and 10 nM. Bose et al. [19] exposed E15 primary cultures of rat embryonic cortical neural stem cells to 2.5 nM and 5 nM MeHg for 48 h and reported reduced cell proliferation with no effect on the cell death rate.

Our new findings on the increase in premature differentiation during embryonic development at a sub-nanomolar MeHg dose might be an outcome of epigenetic changes triggered by stress sensors, such as AMP dependent kinase (AMPK). Even though it is not known whether MeHg can increase AMPK activity, $HgCl_2$ has been shown to enhance AMPK activation in the liver of mice [20]. Hwang et al. [21] have shown that the activation (phosphorylation) of AMPK can play an important role in reducing the toxicity of methylmercury. Therefore, the activation of AMPK could be a biological response induced by the extremely low dose of MeHg. Our previous work has shown that AMPK activation can stimulate a signaling-directed epigenetic pathway, atypical protein kinase C (aPKC)-mediated S436 phosphorylation of CREB-binding protein and histone acetyltransferase, to promote the neuronal differentiation of embryonic and adult neural precursor cells [22,23]. Moreover, it has been reported that the levels of cysteine and glutathione (GSH) as well as the GSH/GSSG ratio in neural stem cells progressively decreased in association with neuronal differentiation [24]. Since it is well known that MeHg decreases GSH, this may also be a potential mechanism for the observed effects.

This enhanced neuronal differentiation by sub-nanomolar MeHg could have significant biological consequences. The untimely enhancement of embryonic neurogenesis can lead to depletion of the neural precursor cell pool and consequently a decreased level of adult neurogenesis resulting in neurological functional impairment. Juliandi et al. [25] have shown that prenatal treatment of valproic acid in mice can enhance neurogenesis and reduce the proliferation of neural precursor cells (NPCs), leading to the depletion of the NPC pool. This depletion may cause a slower differentiation of the residual NPCs during life. In contrast, Gallaher et al. [26] showed that a maternal IL-6 surge aberrantly affected embryonic precursors, ultimately causing an expanded adult forebrain NPC pool and enhanced olfactory neurogenesis in offspring, months after fetal exposure. The possibility that different doses of prenatal exposure to MeHg can have different effects on the adult NPC pool and its associated neurological effects is an interesting avenue to be explored in the future.

While the doses used in this in-vitro experiment cannot be extrapolated to MeHg doses during human fetal development, the difference of effects observed between the sub-nanomolar range and the nanomolar range has biological significance. An analysis of autopsied brain tissue from infants prenatally exposed to methylmercury showed that the mercury levels detected were 0.026–0.295 µg/g [27]. Sakamoto et al. [28] measured total Hg concentrations in the cord blood of 54 healthy Japanese pregnant women, with no particular exposure to any Hg compounds at Fukuda Hospital (Kumamoto City, Kumamoto, Japan) from 2006 to 2007, and reported a mean Hg concentration of 7.26 ng/g. This means that the nM range of exposure is environmentally relevant.

In conclusion, based on our results, we propose a novel model for low-dose MeHg exposure to neural precursor cells, as shown in Figure 3. Extremely low-dose MeHg exposure may also have a detrimental effect on embryonic neural development, which may lead to neurodevelopmental disorders or neurodegeneration later in life.

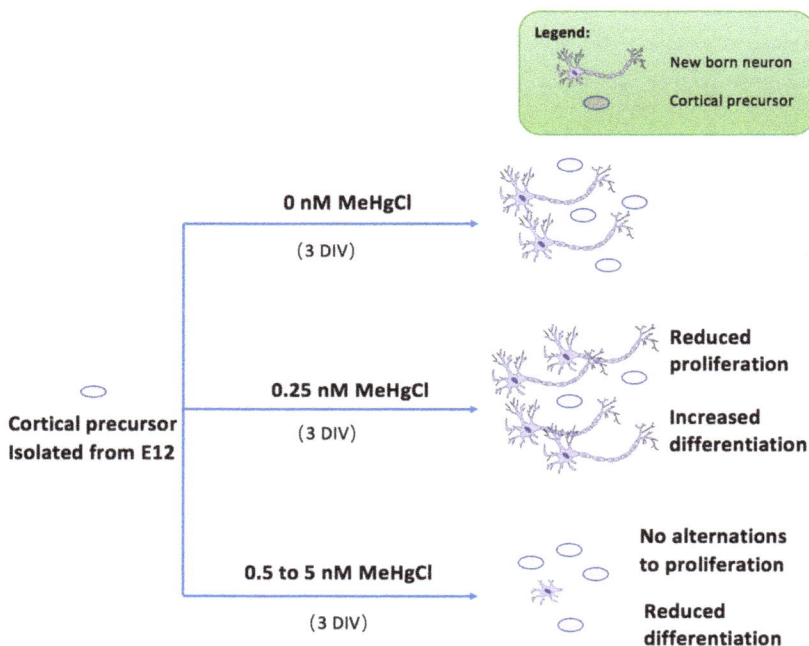

Figure 3. Proposed model for low-dose MeHg exposure to NSCs. Under normal circumstances, cortical precursor cells undergo both differentiation and proliferation. Upon administration of 0.25 nM MeHgCl, the cell population show reduced proliferation and increased differentiation. Cell population dosed with 0.5 nM to 5 nM MeHgCl show inhibited differentiation and gradual recovery of proliferation.

Author Contributions: Conceptualization, H.M.C. and J.W.; methodology, H.M.C. and J.W.; software, X.Y.; validation, H.M.C. and J.W.; formal analysis, X.Y.; investigation, J.W.; resources, H.M.C. and J.W.; data curation, H.M.C. and J.W.; writing-original draft preparation, X.Y.; writing-review and editing, H.M.C. and J.W.; visualization, X.Y.; supervision, H.M.C. and J.W.; project administration, H.M.C. and J.W.; and funding acquisition, H.M.C.

Funding: This research was funded by NSERC Discovery Grant and Canada Research Chair Grant to H.M.C., X.Y. received a NSERC CREATE-Research in Environmental and Analytical Chemistry.

Acknowledgments: We thank Charvi Syal, Sarma Nath Sailendra, Matt Seegobin, Jayasankar Kosaraju, and Emmanuel Yumvihoze for their technical support.

Conflicts of Interest: The authors declare no conflicts of interest. The funders had no role in the design of the study; in the collection, analyses, or interpretation of data; in the writing of the manuscript, and in the decision to publish the results.

References

1. Mergler, D.; Anderson, H.A.; Chan, L.H.M.; Mahaffey, K.R.; Murray, M.; Sakamoto, M.; Stern, A.H. Methylmercury exposure and health effects in humans: A worldwide concern. *AMBIO J. Hum. Environ.* **2007**, *36*, 3–11. [CrossRef]

2. Driscoll, C.T.; Mason, R.P.; Chan, H.M.; Jacob, D.J.; Pirrone, N. Mercury as a global pollutant: Sources, pathways, and effects. *Environ. Sci. Technol.* **2013**, *47*, 4967–4983. [CrossRef] [PubMed]

3. Ha, E.; Basu, N.; Bose-O'Reilly, S.; Dórea, J.G.; McSorley, E.; Sakamoto, M.; Chan, H.M. Current progress on understanding the impact of mercury on human health. *Environ. Res.* **2017**, *152*, 419–433. [CrossRef] [PubMed]

4. Amin Zaki, L.; Elhassani, S.; Majeed, M.A.; Clarkson, T.W.; Doherty, R.A.; Greenwood, M.R.; Giovanoli-Jakubczak, T. Perinatal Methylmercury Poisoning in Iraq. *Am. J. Dis. Child.* **1976**, *130*, 1070–1076. [CrossRef] [PubMed]

5. Matsumoto, H.; Koya, G.; Takeuchi, T. Fetal Minamata disease: A neuropathological study of two cases of intrauterine intoxication by a methyl mercury compound. *J. Neuropathol. Exp. Neurol.* **1965**, *24*, 563–574. [CrossRef] [PubMed]

6. Debes, F.; Budtz-Jørgensen, E.; Weihe, P.; White, R.F.; Grandjean, P. Impact of prenatal methylmercury exposure on neurobehavioral function at age 14 years. *Neurotoxicol. Teratol.* **2006**, *28*, 536–547. [CrossRef] [PubMed]

7. Yorifuji, T.; Tsuda, T.; Inoue, S.; Takao, S.; Harada, M. Long-term exposure to methylmercury and psychiatric symptoms in residents of Minamata, Japan. *Environ. Int.* **2011**, *37*, 907–913. [CrossRef] [PubMed]

8. Onishchenko, N.; Tamm, C.; Vahter, M.; Hökfelt, T.; Johnson, J.A.; Johnson, D.A.; Ceccatelli, S. Developmental exposure to methylmercury alters learning and induces depression-like behavior in male mice. *Toxicol. Sci.* **2007**, *97*, 428–437. [CrossRef] [PubMed]

9. Onishchenko, N.; Karpova, N.; Sabri, F.; Castrén, E.; Ceccatelli, S. Long-lasting depression-like behavior and epigenetic changes of BDNF gene expression induced by perinatal exposure to methylmercury. *J. Neurochem.* **2008**, *106*, 1378–1387. [CrossRef] [PubMed]

10. Ferraro, L.; Tomasini, M.C.; Tanganelli, S.; Mazza, R.; Coluccia, A.; Carratù, M.R.; Antonelli, T. Developmental exposure to methylmercury elicits early cell death in the cerebral cortex and long-term memory deficits in the rat. *Int. J. Dev. Neurosci.* **2009**, *27*, 165–174. [CrossRef] [PubMed]

11. Xu, M.; Yan, C.; Tian, Y.; Yuan, X.; Shen, X. Effects of low level of methylmercury on proliferation of cortical progenitor cells. *Brain Res.* **2010**, *1359*, 272–280. [CrossRef] [PubMed]

12. Tamm, C.; Duckworth, J.; Hermanson, O.; Ceccatelli, S. High susceptibility of neural stem cells to methylmercury toxicity: Effects on cell survival and neuronal differentiation. *J. Neurochem.* **2006**, *97*, 69–78. [CrossRef] [PubMed]

13. Theunissen, P.T.; Pennings, J.L.A.; Robinson, J.F.; Claessen, S.M.H.; Kleinjans, J.C.S.; Piersma, A.H. Time-response evaluation by transcriptomics of methylmercury effects on neural differentiation of murine embryonic stem cells. *Toxicol Sci.* **2011**, *122*, 437–447. [CrossRef] [PubMed]

14. Burke, K.; Cheng, Y.; Li, B.; Petrov, A.; Joshi, P.; Berman, R.F.; Reuhl, K.R.; DiCicco-Bloom, E. Methylmercury elicits rapid inhibition of cell proliferation in the developing brain and decreases cell cycle regulator, cyclin E. *Neurotoxicology* **2006**, *27*, 970–981. [CrossRef] [PubMed]

15. Silveira, P.P.; Portella, A.K.; Goldani, M.Z.; Barbieri, M.A. Developmental origins of health and disease (DOHaD). *J. Pediatr.* **2007**, *83*, 494–504. [CrossRef]

16. Gillman, M.W.; Barker, D.; Bier, D.; Cagampang, F.; Challis, J.; Fall, C.; Thornburg, K.L. Meeting Report on the 3rd International Congress on Developmental Origins of Health and Disease (DOHaD). *Pediatr. Res.* **2007**, *61*, 625–629. [CrossRef] [PubMed]

17. Fujimura, M.; Usuki, F. Low concentrations of methylmercury inhibit neural progenitor cell proliferation associated with up-regulation of glycogen synthase kinase 3β and subsequent degradation of cyclin E in rats. *Toxicol Appl Pharmacol.* **2015**, *288*, 19–25. [CrossRef] [PubMed]

18. Tamm, C.; Duckworth, J.K.; Hermanson, O.; Ceccatelli, S. Methylmercury inhibits differentiation of rat neural stem cells via Notch signaling. *NeuroReport* **2008**, *19*, 339–343. [CrossRef] [PubMed]

19. Bose, R.; Onishchenko, N.; Edoff, K.; Lang, A.M.J.; Ceccatelli, S. Inherited effects of low-dose exposure to methylmercury in neural stem cells. *Toxicol. Sci.* **2012**, *130*, 383–390. [CrossRef] [PubMed]

20. Kawakami, T.; Hanao, N.; Nishiyama, K.; Kadota, Y.; Inoue, M.; Sato, M.; Suzuki, S. Differential effects of cobalt and mercury on lipid metabolism in the white adipose tissue of high-fat diet-induced obesity mice. *Toxicol. Appl. Pharmacol.* **2012**, *258*, 32–42. [CrossRef] [PubMed]

21. Hwang, G.W.; Tobita, M.; Takahashi, T.; Kuge, S.; Kita, K.; Naganuma, A. siRNA-mediated AMPKalpha1 subunit gene PRKAA1 silencing enhances methylmercury toxicity in HEK293 cells. *J. Toxicol. Sci.* **2010**, *35*, 601–604. [CrossRef] [PubMed]

22. Wang, J.; Gallagher, D.; DeVito, L.; Cancino, I.; Tsui, D.; He, L.; Keller, G.M.; Frankland, P.W.; Kaplan, D.R.; Miller, F.D. Metformin activates atypical PKC-CBP pathway to promote neurogenesis and enhance spatial memory formation. *Cell Stem Cell* **2012**, *11*, 23–35. [CrossRef] [PubMed]

23. Fatt, M.; Hsu, K.; He, L.; Wondisford, F.; Miller, F.D.; Kaplan, D.R.; Wang, J. Metformin acts on two different molecular pathways to enhance adult neural precursor proliferation/self-renewal and differentiation. *Stem Cell Reports* **2015**, *5*, 988–995. [CrossRef] [PubMed]

24. Trivedi, M.; Zhang, Y.; Lopez-Toledano, M.; Clarke, A.; Deth, R. Differential neurogenic effects of casein-derived opioid peptides on neuronal stem cells: Implications for redox-based epigenetic changes. *J. Nutr. Biochem.* **2017**, *37*, 39–46. [CrossRef] [PubMed]

25. Juliandi, B.; Tanemura, K.; Igarashi, K.; Tominaga, T.; Furukawa, Y.; Otsuka, M.; Moriyama, N.; Ikegami, D.; Abematsu, M.; Sanosaka, T.; et al. Reduced Adult Hippocampal Neurogenesis and Cognitive Impairments following Prenatal Treatment of the Antiepileptic Drug Valproic Acid. *Stem Cell Reports* **2015**, *5*, 996–1009. [CrossRef] [PubMed]

26. Gallagher, D.; Norman, A.A.; Woodard, C.L.; Yang, G.; Gauthier-Fisher, A.; Fujitani, M.; Vessey, J.P.; Cancino, G.I.; Sachewsky, N.; Woltjen, K.; et al. Transient maternal IL-6 mediates long-lasting changes in neural stem cell pools by deregulating an endogenous self-renewal pathway. *Cell Stem Cell* **2013**, *13*, 564–576. [CrossRef] [PubMed]

27. Lapham, L.W.; Cernichiari, E.; Cox, C.; Myers, G.J.; Baggs, R.B.; Brewer, R.; Shamlaye, C.F.; Davidson, P.W.; Clarkson, T.W. An analysis of autopsy brain tissue from infants prenatally exposed to methylmercury. *Neurotoxicology* **1995**, *16*, 689–704. [PubMed]

28. Sakamoto, M.; Chan, H.M.; Domingo, J.L.; Koriyama, C.; Murata, K. Placental transfer and levels of mercury, selenium, vitamin E, and docosahexaenoic acid in maternal and umbilical cord blood. *Environ. Int.* **2018**, *111*, 309–315. [CrossRef] [PubMed]

Article

Survey of the Extent of the Persisting Effects of Methylmercury Pollution on the Inhabitants around the Shiranui Sea, Japan

Shigeru Takaoka [1],*, Tadashi Fujino [2], Yoshinobu Kawakami [3], Shin-ichi Shigeoka [3] and Takashi Yorifuji [4]

[1]   Kyoritsu Neurology and Rehabilitation Clinic, 2-2-28 Sakurai-cho, Minamata 867-0045, Japan
[2]   Kikuyou Hospital, 5587 Haramizu, Kikuyou 869-1102, Japan; tds-fujino@jcom.zaq.ne.jp
[3]   Minamata Kyoritsu Hospital, 2-2-12 Sakurai-cho, Minamata 867-0045, Japan; mkkawa@fsinet.or.jp (Y.K.);
      shigeoka@mk-kyouritu.com (S.-i.S.)
[4]   Department of Human Ecology, Graduate School of Environmental and Life Science, Okayama University,
      3-1-1 Tsushima-naka, Kita-ku, Okayama 700-8530, Japan; yorichan@md.okayama-u.ac.jp
*   Correspondence: stakaoka@x.email.ne.jp; Tel.: +81-966-63-6835

Received: 28 June 2018; Accepted: 18 July 2018; Published: 20 July 2018

Abstract: In 1956 methylmercury poisoning, known as Minamata disease, was discovered among the inhabitants around the Shiranui Sea, Kyushu, Japan. Although about five hundred thousand people living in the area had supposedly been exposed to methylmercury, administrative agencies and research institutes had not performed any subsequent large scale, continuous health examination, so the actual extent of the negative health effects was not clearly documented. In 2009, we performed health surveys in order to examine residents in the polluted area and to research the extent of the polluted area and period of pollution. We analyzed data collected on 973 people (age = 62.3 ± 11.7) who had lived in the polluted area and had eaten the fish there and a control group, consisting of 142 persons (age = 62.0 ± 10.5), most of whom had not lived in the polluted area. Symptoms and neurological signs were statistically more prevalent in the four groups than in the control group and were more prevalent and severe in those who had eaten most fish. The patterns of positive findings of symptoms and neurological findings in the four groups were similar. Our data indicates that Minamata disease had spread outside of the central area and could still be observed recently, almost 50 years after the Chisso Company's factory had halted the dumping of mercury polluted waste water back in 1968.

Keywords: methylmercury; long term exposure; symptoms; neurological findings; severity; delayed toxicity; correlation of signs and symptoms; dose-response relationship

1. Introduction

In 1956, methylmercury poisoning was discovered among the inhabitants around Minamata Bay of Shiranui Sea in Kumamoto Prefecture, Kyushu, Japan. The condition, which was caused by the ingestion of fish and shellfish that had been contaminated by methylmercury, became known as Minamata disease [1]. For 36 years, from 1932 to 1968, the Nihon Chisso Company's Minamata factory, which produced acetaldehyde, had been discharging waste water contaminated with methylmercury, created during the process, into the Shiranui Sea. Although it was suspected that about five hundred thousand people living in the area had been exposed to methylmercury poisoning, there have subsequently been very few comprehensive health surveys and examinations.

Certification of Minamata disease (methylmercury poisoning by eating fish) has been determined by the Pollution-Related Health Damage Compensation Law (PHDCL). Those victims must themselves personally apply to the Judgment Committee for Minamata Disease Accreditation (an advisory body to the Governor of the Kumamoto Prefecture) to be certified. According to the 1977 Diagnostic Criteria of Minamata disease, patients must have lived in the designated area at least one year before 1968, and four-limb somatosensory disturbance must be accompanied by at least one of the more severe symptoms of Hunter Russell syndrome such as ataxia, visual field constriction, and so on [2].

Actual judgment is stricter. A report by a neurologists in 1997 suggested that patients with such plural symptoms of Hunter Russell syndrome had been rejected by the Judgment Committee [3].

The designated area has not been determined by epidemiological study but restricted to the place where severe Minamata disease victims have been discovered since the outbreak. Therefore, strangely enough, there is an enclave (Akasegawa district) in Akune City (Figure 1). Originally, the designated area was restricted to Minamata City, Tsunagi Town, Ashikita Town (including ex-Tanoura Town), and ex-Izumi City area by the PHDCL in 1971. However, it later became clear that the effects were more outspread and the extent of the area was expanded.

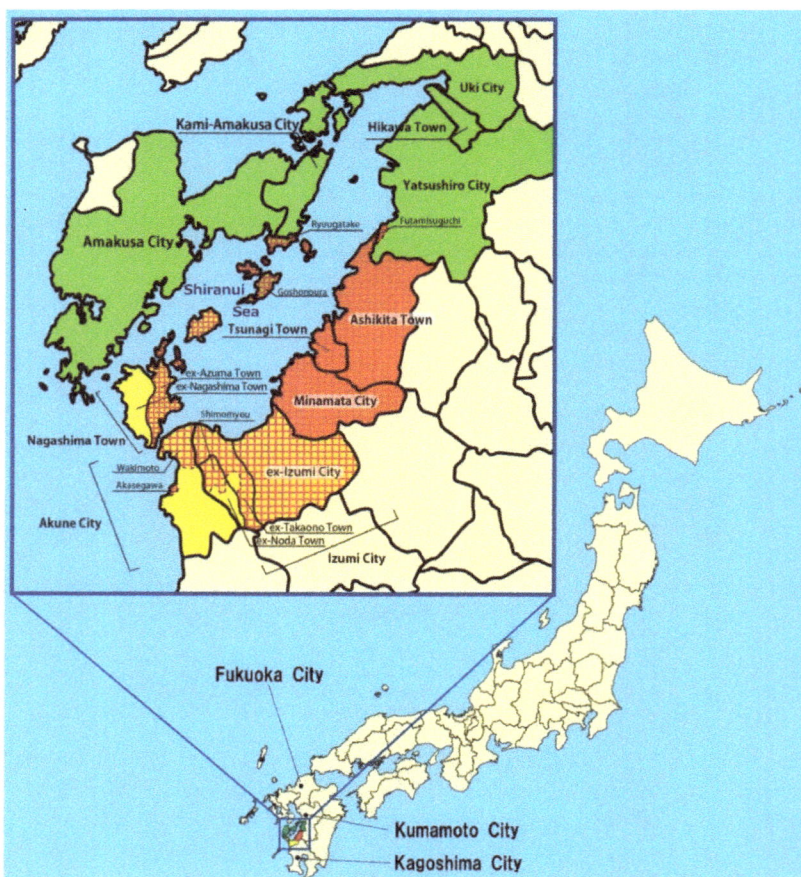

Figure 1. Surveyed area and designated area. Minamata Area: red, Northern Area: green, Southern Area: yellow, Other Areas: beige. Designated Area: red cross-hatched pattern. Control area: around Fukuoka, Kumamoto, and Kagoshima City.

As of 2009, when time-limited special relief law (The Law Concerning Special Measures for the Relief of Minamata Disease Victims and the Settlement of Minamata Disease Issues, LSRS) was enforced, the designated area was restricted to the red cross-hatched areas in Figure 1, which consists of Minamata City, Tsunagi Town, Ashikita Town, Goshonoura of Amakusa City, Ryuugatake of Kami-Amakusa City, Futamisuguchi of Yatsushiro City in Kumamoto Prefecture and Izumi City, ex-Azuma Town area of Nagashima Town, Wakimoto and Akasegawa of Akune City in Kagoshima Prefecture (Table 1). Strictly speaking, the area designated by the PHDCL and that by LSRS are different. The designated area we refer to in this paper is the one stipulated by the Special Law (LSRS) that was in force between 2009 and 31 July 2012.

Table 1. Surveyed area and designated area.

	Designated Area	**Non-Designated Area**
Control Area		around Fukuoka City
		around Kumamoto City
		around Kagoshima City
Minamata Area (Kumamoto Prefecture)	Minamata City	
	Tsunagi Town	
	Ashikita Town *	
Northern Area (Kumamoto Prefecture)	Goshonoura district, Amakusa City **	Other districts of Amakusa City **
	Ryuugatake district, Kami-Amakusa City ***	Other districts of Kami-Amakusa City ***
	Futamisuguchi district, Yatsushiro City	Other districts of Yatsushiro City
		Uki City
		Hikawa Town
Southern Area (Kagoshima Prefecture)	All ex-Izumi City district, Izumi City ****	Other districts of Izumi City ****
	Euchi, Ookubo, Kamizuru, Shimozuru, and Shibabiki of ex-Takaono Town district, Izumi City ****	
	Shimomyo district of ex-Noda Town, Izumi City ****	
	All ex-Azuma Town district, Nagashima Town *****	Other districts (All ex-Nagashima Town) of Nagashima Town *****
	Wakimoto and Akasegawa districts of Akune City	Other districts of Akune City
Other Areas		Other districts of Kumamoto Prefecture
		Other districts of Kagoshima Prefecture
		Other Prefectures

* On 1 January 2005, ex-Ashikita Town and ex-Tanoura Town merged and became Ashikita Town; ** On 27 March 2006, ex-Goshonoura Town, 2 cities, and 7 other towns merged and became Amakusa City; *** On 31 March 2004, ex-Ryuugatake Town and 3 other towns merged and became Kami-Amakusa City; **** On 13 March 2006, ex-Izumi City, ex-Noda Town, and ex-Takaono Town merged and became Izumi City; ***** On 20 March 2006, ex-Azuma Town and ex-Nagashima Town merged and became Nagashima Town.

As of 2007, only 2268 patients were accredited by the administration [4] as suffering from Minamata disease, despite the fact that more than 17,000 people had applied for Minamata disease certification prior to 1999 [5]. In 1995, 8831 individuals were partially compensated for health problems, which included somatosensory disturbance [4], but they were not certified as actually having Minamata disease. Applicants seeking recognition were being socially discriminated against and the lack of a comprehensive pollution survey meant that many residents with health problems had not sought diagnosis.

However, in October 2004, after the Supreme Court of Japan ruled that the criteria stipulated by the Japanese Environmental Protection Agency for Minamata disease accreditation was too strict, an increasing number of residents applied for examination, therapy, and treatment for methylmercury poisoning. By the summer of 2009 over thirty thousand had applied and by the end of 2012 the number had risen to sixty thousand. In order to study the signs and symptoms of residents who hoped to be certified, we carried out a survey in September 2009 to determine the geographical extent and chronological development of methylmercury poisoning in the area.

From November 2004 to August 2009, we had already examined 3800 residents in the polluted area and found a lot of patients with neurological signs and symptoms. We performed this survey in order to research the prevalence of signs and symptoms as well as the geometrical and chronological spread of health problems caused by methylmercury.

In this study, we investigated not only the state of health (symptoms and neurological signs) in methylmercury-exposed people at present but the extent and mutual relationship of exposure level, onset and course of symptoms, severities of signs and symptoms and the significance of the designated area in the polluted region.

Also, the damage to health from methylmercury poisoning has been so great in Japan that researchers have concentrated on finding severe neurological abnormalities and have ignored milder health disturbances. Therefore, we also analyzed the data from subjects in whom sensory disturbances had not been detected during their physical examination.

2. Materials and Methods

2.1. Subjects

The study, which was planned for people who had lived in the methylmercury-polluted area and hoped to be certified for Minamata disease, was carried out on 20 and 21 September 2009. Information regarding the study was spread through media such as television, newspapers, and local government pamphlets. We received applications from people living throughout the coastal regions of the Shiranui Sea. Of the 1700 applicants we selected the first 1420 and informed them of the times, dates and places for the examinations.

Examinations were conducted at twelve sites in Kumamoto Prefecture and at five sites in Kagoshima Prefecture. A total of 1044 subjects were examined. They were given both written and verbal information about the examination method, how the data would be used and that their confidentiality would be protected. Of the 1044 subjects examined 974 residents agreed to having their data analyzed. The Ethics Committee of Minamata Kyoritsu Hospital approved the implement and analyses of this study. Among 974 subjects, one subject had not lived in the polluted area, so we excluded this subject. Finally, we analyzed 973 residents (M/F = 482/491, age = 62.3 $\pm$ 11.7, range = 33–92).

We grouped the exposed subjects into four categories, according to their living places at the time of the examination. Subjects who lived in Minamata City or Ashikita County (Tsunagi Town and Tanoura Town) at the examination time were classified as Minamata Area (n = 259, M/F = 136/123, Age = 62.5 $\pm$ 13.6). Subjects who lived in the northern area including Amakusa City, Kami-Amakusa City, Yatsushiro City, Yatsushiro County (Hikawa Town), or Uki City were classified as Northern Area (n = 279, M/F = 147/132, Age = 64.0 $\pm$ 11.1). Subjects who lived in Izumi City, Izumi County (Nagashima Town), or Akune City were classified as Southern Area (n = 246, M/F = 121/125, Age = 63.0 $\pm$ 11.1). Subjects who lived in the polluted area from 1950s to 1970s and were now living in other areas of Japan at the time of the examination were classified as Other Area (n = 189, M/F = 88/101, Age = 58.6 $\pm$ 9.9) (Tables 1 and 2, Figure 1).

The Control group was comprised of 227 persons chosen from hospital staff and residents living around Fukuoka City, Kumamoto City, and Kagoshima City in 2006 and 2007. In this control group, younger subjects were over-represented, and were excluded. Finally, we selected 142 residents (n = 142, M/F = 56/86, age = 62.0 $\pm$ 10.5, range = 36–86). Most of them had worked in the service sector. The subjects in the control group were the same as those mentioned in another paper in 2008 [6].

Table 2. Demographic characteristics of subjects in each area (*n* = 1115).

	Control Area	Minamata Area	Northern Area	Southern Area	Other Areas
	(*n* = 142)	(*n* = 259)	(*n* = 279)	(*n* = 246)	(*n* = 189)
Sex, *n* (%)					
Male	56 (39.4)	136 (52.5)	147 (52.7)	121 (49.2)	88 (46.6)
Female	86 (60.6)	123 (47.5)	132 (47.3)	125 (50.8)	101 (53.4)
Age					
Mean ± SD	62.0 ± 10.5	62.5 ± 13.6	64.0 ± 11.1	63.0 ± 11.1	58.6 ± 9.9
Range (min–max)	36–86	33–92	33–90	36–88	34–82
Residential history in designated area (DA) more than 1 year, *n* (%)					
In DA > 1 year (DA)	3 (2.1)	234 (90.3)	156 (55.9)	222 (90.2)	174 (92.1)
Not in DA > 1 year (NDA)	139 (97.9)	12 (4.6)	116 (41.6)	19 (7.7)	10 (5.3)
Born after 1968 (BA1968)	0 (0.0)	13 (5.0)	7 (2.5)	5 (2.0)	5 (2.6)
Smoking, *n* (%)					
Non-smoker	109 (77.3)	198 (76.4)	226 (81.0)	185 (75.2)	133 (70.4)
Smoker	32 (22.7)	61 (23.6)	53 (19.0)	61 (24.8)	56 (29.6)
Alcohol drinking, *n* (%)					
Non-drinker	72 (51.1)	136 (52.5)	148 (53.0)	129 (52.4)	98 (51.9)
Drinker	69 (48.9)	123 (47.5)	131 (47.0)	117 (47.6)	91 (48.1)
Frequency of fish intake, *n* (%)					
Three times a day	6 (4.4)	95 (36.7)	166 (59.5)	99 (40.2)	99 (52.4)
Twice a day	7 (5.1)	61 (23.6)	70 (25.1)	94 (38.2)	46 (24.3)
Once a day	27 (19.9)	59 (22.8)	27 (9.7)	35 (14.2)	25 (13.2)
More than once a week	63 (46.3)	35 (13.5)	13 (4.7)	16 (6.5)	15 (7.9)
Less than once a week	33 (24.3)	9 (3.5)	3 (1.1)	2 (0.8)	4 (2.1)
Occupation, *n* (%)					
Fishermen (subject)	0 (0.0)	7 (2.7)	82 (29.4)	32 (13.0)	10 (5.3)
Fishermen (subject's parent)	2 (1.4)	28 (10.8)	153 (54.8)	78 (31.7)	60 (31.7)
Complications, *n* (%)					
Hypertension	40 (28.2)	85 (32.8)	118 (42.3)	102 (41.5)	43 (22.8)
Renal diseases	3 (2.1)	17 (6.6)	10 (3.6)	13 (5.3)	15 (7.9)
Liver diseases	6 (4.2)	20 (7.7)	17 (6.1)	20 (8.1)	14 (7.4)
Respiratory diseases	12 (8.5)	14 (5.4)	8 (2.9)	6 (2.4)	13 (6.9)
Diabetes Mellitus	3 (2.1)	17 (6.6)	37 (13.3)	24 (9.8)	15 (7.9)
Orthopedic diseases	13 (9.2)	60 (23.2)	72 (25.8)	62 (25.2)	52 (27.5)
Malignant diseases	7 (4.9)	12 (4.6)	11 (3.9)	17 (6.9)	11 (5.8)
History of application for Minamata disease, *n* (%)	0 (0.0)	33 (12.7)	24 (8.6)	38 (15.4)	17 (9.0)
Family history of Minamata disease, *n* (%)	0 (0.0)	164 (63.3)	108 (38.7)	123 (50.0)	151 (79.9)
Have witnessed abnormal animal behavior, *n* (%)	No Data	100 (38.6)	103 (36.9)	108 (43.9)	76 (40.2)

2.2. Epidemiological Conditions

Due to the absence of a comprehensive pollution survey in the Minamata area, most of the residents have not undergone any tests to determine the mercury levels in their bodies resulting from their exposure to mercury and methylmercury. Therefore, we prepared a questionnaire to collect information about their places of residence, dietary habits, occupations and family health. Such information is important and useful, because, although a large part of the local diet consists of fish, there have been no previous surveys done to gather accurate information. All subjects had eaten fish and shellfish from the Shiranui Sea. In order to assess the indirect methylmercury exposure, when subjects were asked about the frequency of fish ingestion they were asked to choose from one of the five answers (3 times/day, 2 times/day, once/day, more than once/week, less than once/week).

According to their residential history, subjects were classified into one of the following three categories. The first category consisted of subjects who had lived in the designated area for at least one year (DA: n = 786, M/F = 386/400, Age = 63.0 ± 11.3). The second category consisted of subjects who had not lived in the designated area for at least one year (NDA: n = 158, M/F = 85/72, Age = 63.6 ± 10.0). Those who had been born or had moved to the polluted on or after 1 January 1969 were classified under the third category (BA1968: n = 30, M/F = 21/9, Age = 37.4 ± 2.3).

In order to analyze the younger generation, we re-analyzed 30 subjects who were born after 31 December 1968. To evaluate this group, we selected 88 out of 227 subjects whose age was lower than 49 from the Control Area (M/F = 40/48, Age = 37.5 ± 6.0), and 84 out of 786 exposed subjects in the designated area who were born after 31 December 1968 (M/F = 44/40, Age = 44.8 ± 2.3) and whose age was lower than 49 from the four exposed groups.

The subjects' age, sex, smoking and alcohol drinking habits, frequency of fish intake, the occupation of the subjects and the subjects' parents, complications, history of application for Minamata disease, family history of Minamata disease, and the witnessing of abnormal animal behavior was analyzed for each group.

2.3. Onset of Symptoms

Subjects from the four polluted areas were asked about the time of onset of their first abnormal experience supposed to be related to methylmercury exposure, muscle cramps, four-limb numbness, stumbling tendency, difficulty in fine finger tasks, and limited peripheral vision. The time between the appearance of the first symptom until all five symptoms (cramps, numbness, stumbling tendency, difficulty in fine finger tasks, and limited vision) had appeared were calculated. When a subject could not answer with certainty, the approximate year or median of the generation was used.

The latency period between methylmercury exposure and onset of health problems can be roughly estimated by this information.

2.4. Questionnaire on Complaints

Our questionnaire consisted of 57 questions. Seven questions were related to sensory impairment, 6 related to somatic pain, 6 related to visual impairment, 3 related to hearing impairment, 3 related to tasting and smelling problems, 9 related to in-coordination of the extremities, 5 related to other movement impairment, 4 related to vertigo and dizziness, 3 related to general complaints, 11 related to emotional and intellectual problems. The subjects were asked to answer each question by selecting from one of the following four possibilities: (1) Yes, always; (2) Yes, sometimes; (3) Yes in the past but No at present; (4) Never. The prevalence of each complaint was calculated for each group and compared. The subjects completed the questionnaire before they were examined. Any subjects who were unable to complete the questionnaire by themselves were questioned orally. The results of the questionnaires were reviewed prior to the examination.

2.5. Neurological Examination

A neurological examination was performed on all subjects. The examination comprised the following tests: dysarthria, auditory disturbance, visual constriction, gait disturbance, tandem gait, Mann's test, balancing on one foot (eyes open and closed), finger-to-nose test (eyes open and closed), diadochokinesis, heel-to-knee test, postural tremor of hand, and superficial sensory disturbance (touch and pain).

Dysarthria, auditory disturbance, visual constriction, and postural tremor were judged as present or absent. Dysarthria, auditory disturbance, and visual field were judged by the examining physician without using special instruments. Visual disturbance was considered present when the subject had a lateral visual field of 80 degrees or less, as measured by the confrontation visual field test.

Limb and truncal ataxia were judged as absent, mildly abnormal, or distinctly abnormal. Tandem gait disturbance was judged as distinctly abnormal if the subject could not walk more than 5 steps and as mildly abnormal if they could walk 5 steps but were unstable. The one-foot standing and Mann's test were judged as distinctly abnormal if it was impossible for the subject to keep their balance for more than 3 s and as mildly abnormal if they could keep their balance for more than 3 s. Finger-to-nose test and heel-to-knee test were judged as distinctly abnormal if there was constant dysmetria or decomposition and as mildly abnormal if there was uncertain dysmetria, decomposition, or slow progress to the destination. Dysdiadochokinesis was judged as distinctly abnormal if there was a constant abnormality and as mildly abnormal if there was an uncertain abnormality or slow movement.

For sensory disturbance, tactile sense was examined by comparing chest area and dorsal side of both hands and both foot by using a calligraphy brush. After that, truncal and perioral tactile sense was examined by touching the skin softly with the brush. Pain sense was examined by comparing chest area and dorsal side of both hands and both foot by using a 20 g needle for pain inspection. After that truncal and perioral pain sense were examined by the same needle. The needle was attached to a 20 g handle. We evaluated the tactile and pain sense both by relative evaluation between different sites of bodies and by absolute evaluation, especially for pain. When general tactile sensory disturbance was suspected, a physician asked of the subject to close his or her eyes and indicate which part of their body was being touched by the calligraphy brush.

One-hundred and forty-four doctors, including neurological specialists, carried out primary and secondary examinations on the subjects. The sensory examination was repeated by more trained physicians on all subjects.

All the physicians participating in the study were trained in the procedures by direct instruction, written instruction or visual instruction in the form of video tutorials. The physicians who performed the neurological examination in the control areas were different from those in the polluted area, but the methods and criteria of the neurological examination was the same as those preformed in the polluted areas.

2.6. Statistical Methods

As formerly mentioned in Section 2.1 in this chapter, we classified the methylmercury-exposed subjects into four groups: Minamata Area, Northern Area, Southern Area, and Other Areas. We compared them with the Control Area. All the calculations were performed using MS Excel 2010 and STATA ver.14. The prevalence of data was analyzed by using MS Excel and STATA. Logistic regression analysis and correlation analysis were performed by STATA. When the prevalence of the control group was zero in an item of symptoms or signs, we postulated that the eldest subject in the control group was positive and calculated the OR and confidence interval by STATA. The analysis of the onset year was done in MS Excel.

2.6.1. Questionnaire on Symptoms and Neurological Examination

The data percentages of the answers "always yes" and "always or sometimes yes" from the questionnaire were summed, the results analyzed and the correlations between the control and the four exposed groups were calculated. A total of 23 questions out of the 57 questions asked were analyzed. Three of the 7 questions relating to sensory impairment, 2 of the 6 relating to somatic pain, 1 of the 6 relating to visual impairment, 1 of the 3 relating to hearing impairment, 2 of the 3 relating to tasting and smelling problems, 5 of the 9 relating to in-coordination of the extremities, 2 of the 5 relating to other movement impairment, 1 of the 4 relating to vertigo and dizziness, 1 of the 3 relating to general complaints, and 5 of the 11 relating to emotional and intellectual problems were selected and analyzed.

The prevalence of each sign and symptom from the five groups (1 control and 4 exposed groups) were calculated. Odd's ratio for the association between area and symptoms or signs were calculated by logistic regression analysis. Correlation of prevalence among the control group and the four other exposed groups were calculated for symptoms (always "yes"), symptoms (always or sometimes "yes"), and neurological signs.

2.6.2. Score for Symptoms (Always), Symptoms (Always and Sometimes), and Neurological Signs

To evaluate the severity of the neurological signs, we added (a) mark(s) to positive signs and symptoms, and we calculated the total score in the exposed four groups.

As to the symptom score (always), we added 1 point when a subject's answer was "always". Symptoms score (always) ranged from 0 to 23. As to symptoms (always and sometimes), we added 2 points when a subject's answer was "always", and added 1 point when a subject's answer was "sometimes". Symptoms score (always and sometimes) ranged from 0 to 46. A score of zero was given to "no answer" items.

As to the neurological signs, the cranial nerve score (6 points) consists of dysarthria (2), auditory disturbance (2), visual constriction (2). The upper, lower ataxia and tremor score (5 points) consists of finger-to-nose test (eyes open) (1), finger-to-nose test (eyes closed) (1), diadochokinesis (1), heel-to-knee test (1), and postural tremor (1). The truncal ataxia score (5 points) consists of normal gait disturbance (1), tandem gait disturbance (1), Mann's test (1), balancing on one foot (eyes open) (1), and balancing on one foot (eyes closed) (1). To evaluate upper, lower, and truncal ataxia, 1 point was given for both mild and distinct abnormalities. The sensory score (6 points) consists of four-limb peripheral touch disturbance (1), perioral touch disturbance (1), systemic touch disturbance below the neck (1), four-limb peripheral pain disturbance (1), perioral pain disturbance (1), systemic pain disturbance below the neck (1). We finally totaled the four scores (22 points) for cranial nerve (6), upper, lower ataxia and tremor (5), truncal ataxia (5), and sensory (6). A score of zero was given to the "no data" item.

The scores were calculated for each area. To research the presence of dose-response relationship, each score according to the frequency of fish ingestion in the exposed groups were calculated.

2.6.3. Analyses among Fish Ingestion, Score of Signs and Symptoms, and the Onset of Symptoms

In Minamata, most of the residents had not had their mercury levels measured during the most polluted period or in the subsequent period when mercury level had decreased. Instead of such direct mercury pollution values, we used frequency of fish ingestion as an indirect indication of methylmercury exposure.

To estimate dose-response (from methylmercury exposure to health effects) relationships, we calculated scores of signs and symptoms for each frequency of fish ingestion. In order to estimate the difference of latency period from exposure of mercury by exposure levels, we calculated the average year of onset and the average interval between the first symptom and the onset of each following symptom in each frequency of fish ingestion.

Lastly, we evaluated the correlation between the score for signs and symptoms, and onset period of symptoms to estimate difference of latency period and severity of methylmercury poisoning.

2.6.4. Characteristics of Signs and Symptoms in Subjects Whose Sensory Disturbance Was Not Recognized

To estimate the degree of health disturbance in subjects in whom sensory disturbance had not been detected during their physical examination, we re-classified the exposed subjects into two groups. Sensory disturbance ($-$) group (n = 91, M/F = 64/27, Age = 59.7 $\pm$ 13.3) consisted of subjects whose sensory score was zero, in whom neither four-limb peripheral disturbance, perioral disturbance, systemic disturbance below the neck in tactile or pain examination met the criteria required for recognition. Sensory disturbance (+) group consisted of other subjects whose sensory score was one to six (n = 882, M/F = 428/454, Age = 62.6 $\pm$ 11.6). The control group was the same as those used in the first analysis. We compared signs and symptoms between the three groups.

3. Results

3.1. The Subjects' Background

The demographic characteristics of the subjects from the five groups are shown in Table 2. The age for Other Areas was significantly lower than other groups. Smoking and alcohol drinking was almost the same in the different groups. The frequency of fish intake was significantly higher in the exposed groups than the control. Hypertension, diabetes mellitus, and orthopedic diseases were significantly more prevalent in the polluted area. Subjects who had applied for Minamata disease certification were only 8.6% to 15.4%, even in the polluted area, although many (38.7–79.9%) of them had a family history of Minamata disease.

3.2. Questionnaire on Symptoms

The symptoms (always) are shown in Table 3 and Figure 2. For most of the symptoms (always), Odds ratio (OR) for the association between area and signs or symptoms calculated by logistic regression analysis and adjusted for age, sex, diabetes mellitus, and orthopedic diseases were very high and the lower limits of the OR were higher than 1, except for two questions ("perioral numbness" and "swaying or dizziness" between control and Minamata Area). The correlation among the four exposed groups on the prevalence of symptoms (always) (Table 4) were extremely high (0.9246–0.9611) whereas the correlation between the control and the four exposed groups were very low (0.2751–0.3866).

Table 3. Prevalence of symptoms (Always) and adjusted * odds ratios (OR) for the association between area and symptoms (*n* = 1115).

	Control Area	Minamata Area	Northern Area	Southern Area	Other Areas
Sensory numbness in both hands					
case/N (%)	3/138 (2.2)	98/259 (37.8)	116/278 (41.7)	126/246 (51.2)	73/189 (38.6)
OR (95% CI)	1 (reference)	27 (8.3–87)	30 (9.4–98)	46 (14–148)	29 (9.0–96)
Sensory numbness in both legs					
case/N (%)	1/139 (0.7)	98/259 (37.8)	125/279 (44.8)	126/246 (51.2)	68/189 (36.0)
OR (95% CI)	1 (reference)	80 (11–586)	100 (14–731)	140 (19–1022)	86 (12–633)
Perioral numbness **					
case/N (%)	0/107 (0.0)	15/259 (5.8)	25/279 (9.0)	31/246 (12.6)	14/189 (7.4)
OR (95% CI)	1 (reference)	5.8 (0.7–45)	8.8 (1.2–66)	13.9 (1.9–104)	9.6 (1.2–75)
Headache **					
case/N (%)	0/137 (0.0)	44/259 (17.0)	52/279 (18.6)	59/246 (24)	47/189 (24.9)
OR (95% CI)	1 (reference)	31 (4.2–228)	35 (4.7–254)	47 (6.4–344)	49 (6.6–358)
Muscle cramps					
case/N (%)	5/137 (3.6)	62/259 (23.9)	65/279 (23.3)	59/245 (24.1)	42/189 (22.2)
OR (95% CI)	1 (reference)	8.5 (3.3–22)	8.1 (3.2–21)	8.5 (3.3–22)	8.3 (3.2–22)
Limited peripheral vision					
case/N (%)	1/138 (0.7)	67/259 (25.9)	91/279 (32.6)	94/246 (38.2)	59/72 (81.9)
OR (95% CI)	1 (reference)	47 (6.5–347)	63 (8.6–457)	84 (12–614)	71 (9.6–520)
Difficulty in hearing					
case/N (%)	12/137 (8.8)	90/259 (34.7)	106/278 (38.1)	130/246 (52.8)	71/189 (37.6)
OR (95% CI)	1 (reference)	5.1 (2.7–9.9)	5.5 (2.9–11)	11 (5.9–22)	7.2 (3.7–14)
Difficulty in smelling					
case/N (%)	1/139 (0.7)	45/259 (17.4)	58/279 (20.8)	46/245 (18.8)	31/189 (16.4)
OR (95% CI)	1 (reference)	29 (3.9–213)	36 (5.0–266)	32 (4.4–238)	31 (4.1–228)
Difficulty in tasting **					
case/N (%)	0/140 (0.0)	34/259 (13.1)	50/279 (17.9)	46/245 (18.8)	23/189 (12.2)
OR (95% CI)	1 (reference)	20 (2.7–146)	28 (3.9–208)	31 (4.2–226)	20 (2.7–151)

Table 3. *Cont.*

	Control Area	Minamata Area	Northern Area	Southern Area	Other Areas
Stumbling tendency **					
case/N (%)	0/106 (0.0)	72/259 (27.8)	99/279 (35.5)	89/246 (36.2)	48/189 (25.4)
OR (95% CI)	1 (reference)	41 (5.6–304)	59 (8.0–429)	63 (8.6–463)	45 (6.1–336)
Staggering **					
case/N (%)	0/107 (0.0)	57/259 (22.0)	76/278 (27.3)	73/246 (29.7)	33/189 (17.5)
OR (95% CI)	1 (reference)	29 (3.9–212)	38 (5.2–280)	46 (6.2–337)	29 (3.8–215)
Difficulty in fine finger tasks **					
case/N (%)	0/140 (0.0)	94/259 (36.3)	136/279 (48.7)	139/246 (56.5)	87/189 (46.0)
OR (95% CI)	1 (reference)	77 (11–562)	128 (18–930)	182 (25–1326)	135 (18–987)
Difficulty in buttoning **					
case/N (%)	0/140 (0.0)	52/259 (20.1)	73/279 (26.2)	53/245 (21.6)	28/189 (14.8)
OR (95% CI)	1 (reference)	31 (4.2–228)	43 (5.9–319)	36 (4.9–267)	30 (4.0–228)
Dropping things held in the hand **					
case/N (%)	0/140 (0.0)	39/259 (15.1)	60/279 (21.5)	56/246 (22.8)	22/189 (11.6)
OR (95% CI)	1 (reference)	24 (3.2–176)	36 (4.9–266)	41 (5.6–303)	22 (3.0–170)
Difficulty in speaking words or sentences well **					
case/N (%)	0/140 (0.0)	29/259 (11.2)	41/279 (14.7)	41/246 (16.7)	16/189 (8.5)
OR (95% CI)	1 (reference)	17 (2.3–125)	23 (3.1–167)	28 (3.8–206)	15 (2.0–118)
Postural hand tremor					
case/N (%)	2/138 (1.4)	41/259 (15.8)	50/279 (17.9)	48/246 (19.5)	31/189 (16.4)
OR (95% CI)	1 (reference)	12 (2.9–51)	14 (3.2–57)	16 (3.8–67)	16 (3.7–68)
Swaying or dizziness **					
case/N (%)	0/140 (0.0)	10/259 (3.9)	22/279 (7.9)	15/246 (6.1)	14/189 (7.4)
OR (95% CI)	1 (reference)	5.5 (0.7–44)	12 (1.5–88)	9.1 (1.2–70)	13 (1.7–105)
General fatigue					
case/N (%)	1/140 (0.7)	99/259 (38.2)	112/279 (40.1)	131/246 (53.3)	74/189 (39.2)
OR (95% CI)	1 (reference)	90 (12–657)	98 (14–714)	166 (23–1205)	94 (13–686)
Lack of motivation to do things					
case/N (%)	1/140 (0.7)	43/259 (16.6)	55/279 (19.7)	63/246 (25.6)	40/189 (21.2)
OR (95% CI)	1 (reference)	32 (4.4–237)	40 (5.5–294)	56 (7.6–409)	46 (6.2–341)
Losing your train of thought during conversations **					
case/N (%)	0/139 (0.0)	32/259 (12.4)	29/279 (10.4)	44/246 (17.9)	17/189 (9.0)
OR (95% CI)	1 (reference)	20 (2.6–146)	16 (2.1–118)	31 (4.1–225)	15 (2.0–118)
Forgetfulness					
case/N (%)	1/140 (0.7)	71/259 (27.4)	89/279 (31.9)	93/246 (37.8)	47/189 (24.9)
OR (95% CI)	1 (reference)	54 (7.4–396)	65 (8.9–473)	89 (12–651)	57 (7.7–419)
Irritation **					
case/N (%)	0/140 (0.0)	63/259 (24.3)	87/279 (31.2)	74/246 (30.1)	50/189 (26.5)
OR (95% CI)	1 (reference)	46 (6.3–337)	65 (9.0–475)	61 (8.4–446)	51 (6.9–373)
Anxiety **					
case/N (%)	0/106 (0.0)	78/259 (30.1)	77/279 (27.6)	72/246 (29.3)	56/189 (29.6)
OR (95% CI)	1 (reference)	48 (6.6–354)	43 (5.8–312)	46 (6.2–335)	48 (6.5–351)

* Adjusted for age, sex, diabetes mellitus and orthopedic diseases; ** When prevalence of the control group was zero, we postulated that a positive finding was found in the eldest subject in the control group and calculated the OR and 95% confidence interval.

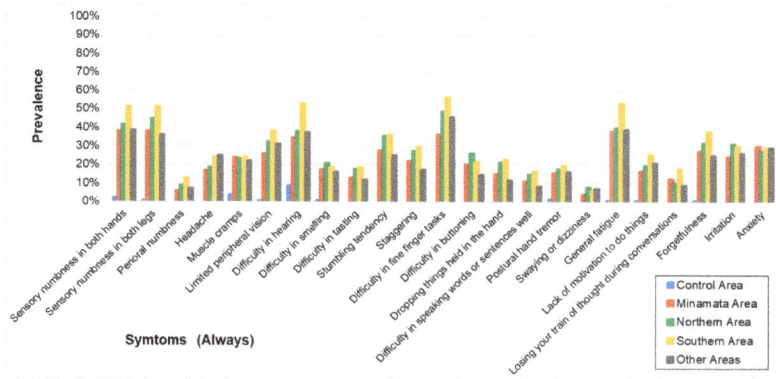

Figure 2. Prevalence of symptoms (Always) in each area. Prevalence was higher in the four exposed groups than that in the control area. The symptomatic patterns were similar among the four groups.

Table 4. Correlation of prevalence of symptoms (Always) among each area.

Correlation/(*p*-value)	Control Area	Minamata Area	Northern Area	Southern Area
Minamata Area (*p*-value)	0.3641 (0.0876)	1.0000		
Northern Area (*p*-value)	0.2751 (0.2040)	0.9611 (0.0000)	1.0000	
Southern Area (*p*-value)	0.3866 (0.0684)	0.9540 (0.0000)	0.9581 (0.0000)	1.0000
Other Areas (*p*-value)	0.3609 (0.0906)	0.9417 (0.0000)	0.9246 (0.0000)	0.9460 (0.0000)

As to the symptoms (always or sometimes), the OR for the same analysis and adjustment were very high and the lower limits of the OR were all higher than 1 (Table 5, Figure 3). The correlation among the four exposed groups on the prevalence of symptoms (always and sometimes) (Table 6) were extremely high (0.9778–0.9875) whereas the correlation between the control and the four exposed groups were lower (0.6501–0.7046).

Table 5. Prevalence of symptoms (Always and Sometimes) and adjusted * odds ratios (OR) for the association between area and symptoms (*n* = 1115).

	Control Area	Minamata Area	Northern Area	Southern Area	Other Areas
Sensory numbness in both hands					
case/N (%)	10/138 (7.2)	221/259 (85.3)	244/278 (87.8)	223/246 (90.7)	160/189 (84.7)
OR (95% CI)	1 (reference)	79 (38–165)	97 (46–206)	130 (59–284)	75 (35–161)
Sensory numbness in both legs					
case/N (%)	10/139 (7.2)	219/259 (84.6)	237/279 (84.9)	215/246 (87.4)	159/189 (84.1)
OR (95% CI)	1 (reference)	77 (37–161)	75 (36–156)	94 (44–201)	81 (37–175)
Perioral numbness **					
case/N (%)	0/107 (0.0)	104/259 (40.2)	136/279 (48.7)	118/246 (48.0)	81/189 (42.9)
OR (95% CI)	1 (reference)	73 (10–533)	104 (14–760)	101 (14–738)	87 (12–636)

Table 5. *Cont.*

	Control Area	Minamata Area	Northern Area	Southern Area	Other Areas
Headache					
case/N (%)	31/137 (22.6)	208/259 (80.3)	219/279 (78.5)	198/246 (80.5)	154/189 (81.5)
OR (95% CI)	1 (reference)	17 (10–29)	16 (9.8–28)	18 (10–30)	17 (9.7–30)
Muscle cramps					
case/N (%)	53/137 (38.7)	237/259 (91.5)	256/279 (91.8)	220/245 (89.8)	172/189 (91.0)
OR (95% CI)	1 (reference)	18 (10–32)	19 (11–33)	15 (8.4–25)	17 (9.2–32)
Limited peripheral vision					
case/N (%)	10/138 (7.2)	171/259 (66.0)	198/279 (71.0)	192/246 (78.0)	132/189 (69.8)
OR (95% CI)	1 (reference)	25 (12–49)	29 (15–59)	44 (22–90)	31 (15–63)
Difficulty in hearing					
case/N (%)	25/137 (18.2)	178/259 (68.7)	212/278 (76.3)	196/246 (79.7)	133/189 (70.4)
OR (95% CI)	1 (reference)	9.8 (5.8–16)	13 (8.0–23)	17 (10–30)	12 (6.8–21)
Difficulty in smelling					
case/N (%)	7/139 (0.5)	124/259 (47.9)	137/279 (49.1)	139/245 (56.7)	96/189 (50.8)
OR (95% CI)	1 (reference)	17 (7.8–39)	18 (8.1–40)	25 (11–55)	20 (8.8–45)
Difficulty in tasting					
case/N (%)	3/140 (2.1)	119/259 (45.9)	148/279 (53.0)	140/245 (57.1)	95/189 (50.3)
OR (95% CI)	1 (reference)	39 (12–126)	51 (16–166)	61 (19–196)	47 (14–152)
Stumbling tendency					
case/N (%)	20/106 (18.9)	213/259 (82.2)	250/279 (89.6)	217/246 (88.2)	165/189 (87.3)
OR (95% CI)	1 (reference)	25 (13–46)	42 (22–81)	37 (19–72)	39 (20–78)
Staggering					
case/N (%)	10/107 (9.3)	197/259 (76.1)	227/278 (81.7)	205/246 (83.3)	149/189 (78.8)
OR (95% CI)	1 (reference)	35 (17–73)	47 (22–98)	53 (25–113)	45 (21–97)
Difficulty in fine finger tasks					
case/N (%)	11/140 (7.9)	195/259 (75.3)	224/279 (80.3)	215/246 (87.4)	148/189 (78.3)
OR (95% CI)	1 (reference)	36 (18–71)	46 (23–92)	81 (39–167)	45 (22–93)
Difficulty in buttoning **					
case/N (%)	0/140 (0.0)	139/259 (53.7)	177/279 (63.4)	149/245 (60.8)	106/189 (56.1)
OR (95% CI)	1 (reference)	158 (22–1147)	224 (31–1629)	211 (29–1536)	199 (27–1458)
Dropping things held in the hand					
case/N (%)	9/140 (6.4)	173/259 (66.8)	212/279 (76.0)	187/246 (76.0)	139/189 (73.5)
OR (95% CI)	1 (reference)	33 (16–68)	51 (24–106)	50 (24–106)	47 (22–100)
Difficulty in speaking words or sentences well					
case/N (%)	4/140 (2.9)	136/259 (52.5)	157/279 (56.3)	154/246 (62.6)	101/189 (53.4)
OR (95% CI)	1 (reference)	39 (14–108)	46 (16–128)	60 (21–167)	43 (15–122)
Postural hand tremor					
case/N (%)	7/138 (5.1)	160/259 (61.8)	188/279 (67.4)	171/246 (69.5)	116/189 (61.4)
OR (95% CI)	1 (reference)	30 (13–66)	37 (17–83)	42 (19–94)	31 (14–70)

Table 5. *Cont.*

	Control Area	Minamata Area	Northern Area	Southern Area	Other Areas
Swaying or dizziness					
case/N (%)	8/140 (5.7)	117/259 (45.2)	157/279 (56.3)	138/246 (56.1)	101/189 (53.4)
OR (95% CI)	1 (reference)	14 (6.6–30)	22 (10–47)	22 (10–47)	20 (9.2–43)
General fatigue					
case/N (%)	30/140 (21.4)	234/259 (90.3)	253/279 (90.7)	230/246 (93.5)	171/189 (90.5)
OR (95% CI)	1 (reference)	43 (23–79)	46 (25–85)	65 (33–128)	37 (19–71)
Lack of motivation to do things					
case/N (%)	31/140 (22.1)	202/259 (78.0)	233/279 (83.5)	215/246 (87.4)	153/189 (81.0)
OR (95% CI)	1 (reference)	15 (9.0–25)	22 (13–38)	30 (17–52)	16 (9.2–28)
Losing your train of thought during conversations					
case/N (%)	11/139 (7.9)	169/259 (65.3)	204/279 (73.1)	183/246 (74.4)	125/189 (66.1)
OR (95% CI)	1 (reference)	24 (12–46)	34 (17–67)	36 (18–72)	26 (13–51)
Forgetfulness					
case/N (%)	79/140 (56.4)	240/259 (92.7)	257/279 (92.1)	235/246 (95.5)	176/189 (93.1)
OR (95% CI)	1 (reference)	11 (6.2–20)	10 (5.4–17)	18 (8.8–37)	13 (6.4–25)
Irritation					
case/N (%)	46/140 (32.9)	204/259 (78.8)	242/279 (86.7)	214/246 (87.0)	164/189 (86.8)
OR (95% CI)	1 (reference)	8.0 (5.0–13)	14 (8.6–24)	14 (8.5–24)	13 (7.6–23)
Anxiety					
case/N (%)	8/106 (7.5)	205/259 (79.2)	225/279 (80.6)	209/246 (85.0)	155/189 (82.0)
OR (95% CI)	1 (reference)	49 (23–109)	55 (25–122)	74 (33–166)	60 (27–136)

* Adjusted for age, sex, diabetes mellitus and orthopedic diseases; ** When prevalence of the control group was zero, we postulated that a positive finding was found in the eldest subject in the control group and calculated the OR and 95% confidence interval.

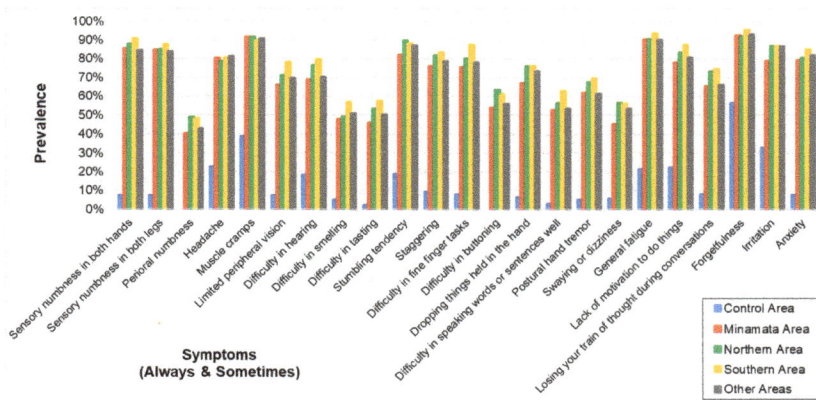

Figure 3. Prevalence of symptoms (Always and Sometimes) in each area. Prevalence was higher in the four exposed groups than that in the control area. The symptomatic patterns were similar among the four groups.

Table 6. Correlation of prevalence of symptoms (Always and Sometimes) among each area.

Correlation/(*p*-Value)	Control Area	Minamata Area	Northern Area	Southern Area
Minamata Area (*p*-value)	0.6921 (0.0003)	1.0000		
Northern Area (*p*-value)	0.6708 (0.0005)	0.9778 (0.0000)	1.0000	
Southern Area (*p*-value)	0.6501 (0.0008)	0.9787 (0.0000)	0.9786 (0.0000)	1.0000
Other Areas (*p*-value)	0.7046 (0.0002)	0.9875 (0.0000)	0.9849 (0.0000)	0.9795 (0.0000)

3.3. Neurological Signs

The results of the neurological signs are shown in Table 7 and Figure 4. Almost all signs were higher in the exposed groups than the control except for the "finger-nose test (opening eyes) (distinct)" in the three polluted areas and "adiadokokinesis (distinct)" in the two polluted areas.

The correlation among the four exposed groups on the prevalence of neurological signs (Table 8) were extremely high (0.9716–0.9864), compared to the correlation between the control and the four exposed groups (0.6327–0.6802).

Table 7. Prevalence of neurological findings and adjusted* odds ratios (OR) for the association between area and neurological findings (*n* = 1115).

	Control Area	Minamata Area	Northern Area	Southern Area	Other Areas
Dysarthria					
case/N (%)	2/141 (1.4)	34/256 (13.3)	37/276 (13.4)	41/242 (16.9)	22/186 (11.8)
OR (95% CI)	1 (reference)	9.7 (2.3–41)	9.3 (2.2–39)	13 (3.1–55)	9.5 (2.2–41)
Hearing loss					
case/N (%)	10/141 (7.1)	59/258 (22.9)	54/279 (19.4)	65/244 (26.6)	27/186 (14.5)
OR (95% CI)	1 (reference)	3.5 (1.7–7.1)	2.6 (1.3–5.4)	4.4 (2.1–9.0)	2.6 (1.2–5.7)
Visual field disturbance **					
case/N (%)	0/142 (0.0)	51/255 (20.0)	66/276 (23.9)	61/243 (25.1)	49/187 (26.2)
OR (95% CI)	1 (reference)	33 (4.4–240)	39 (5.4–288)	44 (6.0–322)	53 (7.2–392)
Normal gait disturbance (distinct) **					
case/N (%)	0/137 (0.0)	24/238 (10.1)	25/255 (9.8)	36/221 (16.3)	15/174 (8.6)
OR (95% CI)	1 (reference)	12 (1.5–87)	11 (1.5–83)	22 (3.0–164)	14 (1.8–107)
Normal gait disturbance (mild-distinct) **					
case/N (%)	0/137 (0.0)	50/238 (21.0)	41/255 (16.1)	52/221 (23.5)	26/174 (14.9)
OR (95% CI)	1 (reference)	30 (4.0–222)	21 (2.8–156)	39 (5.3–292)	32 (4.2–242)
Tandem gait disturbance (distinct)					
case/N (%)	3/141 (2.1)	32/248 (12.9)	41/271 (15.1)	32/241 (13.3)	14/186 (7.5)
OR (95% CI)	1 (reference)	6.0 (1.8–21)	8.0 (2.4–27)	7.2 (2.1–24)	5.7 (1.6–21)
Tandem gait disturbance (mild-distinct)					
case/N (%)	15/141 (10.6)	115/248 (46.4)	151/271 (55.7)	135/241 (56.0)	84/186 (45.2)
OR (95% CI)	1 (reference)	8.0 (4.3–15)	11 (6.0–20)	12 (6.7–23)	10 (5.3–19)
Mann test (distinct)					
case/N (%)	3/109 (2.8)	67/252 (26.6)	76/268 (28.4)	62/242 (25.6)	39/187 (20.9)
OR (95% CI)	1 (reference)	12 (3.7–40)	13 (3.9–42)	12 (3.5–39)	11 (3.3–37)
Mann test (mild-distinct)					
case/N (%)	26/109 (23.9)	153/252 (60.7)	176/268 (65.7)	151/242 (62.4)	114/187 (61.0)
OR (95% CI)	1 (reference)	5.4 (3.2–9.2)	6.2 (3.6–11)	5.6 (3.3–9.5)	6.3 (3.6–11)

Table 7. *Cont.*

	Control Area	Minamata Area	Northern Area	Southern Area	Other Areas
One-foot standing abnormality (eyes open) (distinct)					
case/N (%)	2/141 (1.4)	43/253 (17.0)	61/269 (22.7)	47/242 (19.4)	27/189 (14.3)
OR (95% CI)	1 (reference)	14 (3.3–60)	21 (5.0–89)	18 (4.3–78)	19 (4.4–85)
One-foot standing abnormality (eyes open) (mild-distinct)					
case/N (%)	16/141 (11.3)	121/253 (47.8)	151/269 (56.1)	125/242 (51.7)	85/189 (45.0)
OR (95% CI)	1 (reference)	8.4 (4.6–15)	11 (6.0–20)	9.7 (5.3–18)	9.7 (5.2–18)
One-foot standing abnormality (eyes shut) (distinct)					
case/N (%)	16/142 (11.3)	128/252 (50.8)	135/269 (50.2)	115/243 (47.3)	78/188 (41.5)
OR (95% CI)	1 (reference)	9.7 (5.3–18)	8.4 (4.6–15)	8.0 (4.4–15)	8.2 (4.4–15)
One-foot standing abnormality (eyes shut) (mild-distinct)					
case/N (%)	75/142 (52.8)	202/252 (80.2)	228/269 (84.8)	195/243 (80.2)	150/188 (79.8)
OR (95% CI)	1 (reference)	4.5 (2.7–7.4)	5.1 (3.1–8.5)	3.9 (2.4–6.4)	4.9 (2.9–8.3)
Finger-nose test (eyes open) (distinct) **					
case/N (%)	0/140 (0.0)	10/254 (3.9)	13/264 (4.9)	13/233 (5.6)	11/183 (6.0)
OR (95% CI)	1 (reference)	5.2 (0.6–41)	6.5 (0.8–51)	7.8 (0.99995–61)	9.4 (1.2–75)
Finger-nose test (eyes open) (mild-distinct) **					
case/N (%)	0/140 (0.0)	32/254 (12.6)	61/264 (23.1)	50/233 (21.5)	29/183 (15.8)
OR (95% CI)	1 (reference)	19 (2.6–142)	39 (5.4–288)	37 (5.0–272)	29 (3.8–214)
Finger-nose test (eyes shut) (distinct) **					
case/N (%)	0/139 (0.0)	34/251 (13.5)	31/262 (11.8)	35/233 (15)	28/182 (15.4)
OR (95% CI)	1 (reference)	21 (2.9–157)	18 (2.5–137)	25 (3.4–185)	29 (3.8–215)
Finger-nose test (eyes shut) (mild-distinct)					
case/N (%)	3/139 (2.2)	69/251 (27.5)	95/262 (36.3)	86/233 (36.9)	52/182 (28.6)
OR (95% CI)	1 (reference)	17 (5.1–54)	24 (7.4–78)	26 (7.9–84)	20 (5.9–64)
Adiadokokinesis (distinct) **					
case/N (%)	0/135 (0.0)	11/245 (4.5)	17/270 (6.3)	19/228 (8.3)	15/178 (8.4)
OR (95% CI)	1 (reference)	5.3 (0.7–42)	7.4 (0.96–56)	10.3 (1.4–79)	11.8 (1.5–91)
Adiadokokinesis (mild-distinct)					
case/N (%)	3/135 (2.2)	40/245 (16.3)	85/270 (31.5)	61/228 (26.8)	43/178 (24.2)
OR (95% CI)	1 (reference)	7.9 (2.4–26)	18 (5.6–59)	15 (4.6–49)	16 (4.7–52)
Heel-knee test (distinct) **					
case/N (%)	0/135 (0.0)	17/226 (7.5)	15/239 (6.3)	19/207 (9.2)	9/163 (5.5)
OR (95% CI)	1 (reference)	9.8 (1.3–75)	7.8 (1.0–60)	12 (1.6–93)	8.4 (1.0–68)
Heel-knee test (mild-distinct)					
case/N (%)	3/135 (2.2)	60/226 (26.5)	66/239 (27.6)	68/207 (32.9)	49/163 (30.1)
OR (95% CI)	1 (reference)	16 (4.8–51)	16 (4.9–52)	21 (6.5–70)	22 (6.6–73)
Postural Hand tremor					
case/N (%)	6/136 (4.4)	48/259 (18.5)	58/279 (20.8)	56/246 (22.8)	38/189 (20.1)
OR (95% CI)	1 (reference)	4.7 (2.0–11)	5.3 (2.2–13)	6.1 (2.5–15)	5.6 (2.3–14)
Touch disturbance (four-limb peripheral)					
case/N (%)	1/142 (0.7)	159/259 (61.4)	204/279 (73.1)	164/246 (66.7)	126/189 (66.7)
OR (95% CI)	1 (reference)	218 (30–1581)	368 (50–2678)	271 (37–1977)	265 (36–1944)
Touch disturbance (perioral) **					
case/N (%)	0/142 (0.0)	63/259 (24.3)	71/279 (25.4)	50/246 (20.3)	45/189 (23.8)
OR (95% CI)	1 (reference)	45 (6.2–331)	48 (6.6–353)	36 (4.9–264)	45 (6.1–330)
Touch disturbance (systemic) **					
case/N (%)	0/142 (0.0)	43/259 (16.6)	49/279 (17.6)	43/246 (17.5)	25/189 (13.2)
OR (95% CI)	1 (reference)	28 (3.9–209)	30 (4.1–223)	31 (4.2–225)	24 (3.2–179)

Table 7. *Cont.*

	Control Area	Minamata Area	Northern Area	Southern Area	Other Areas
Pain disturbance (four-limb peripheral)					
case/N (%)	2/142 (1.4)	185/259 (71.4)	224/279 (80.3)	167/246 (67.9)	136/189 (72.0)
OR (95% CI)	1 (reference)	183 (44–760)	296 (71–1238)	153 (37–636)	183 (44–767)
Pain disturbance (perioral) **					
case/N (%)	0/142 (0.0)	103/259 (39.8)	104/279 (37.3)	65/246 (26.4)	60/189 (31.7)
OR (95% CI)	1 (reference)	93 (13–673)	83 (11–603)	50 (6.9–367)	67 (9.1–491)
Pain disturbance (systemic) **					
case/N (%)	0/142 (0.0)	56/259 (21.6)	73/279 (26.2)	56/246 (22.8)	38/189 (20.1)
OR (95% CI)	1 (reference)	38 (5.2–281)	49 (6.7–355)	41 (5.6–300)	37 (5.0–273)

* Adjusted for age, sex, diabetes mellitus and orthopedic diseases; ** When prevalence of the control group was zero, we postulated that a positive finding was found in the eldest subject in the control group and calculated the OR and 95% confidence interval.

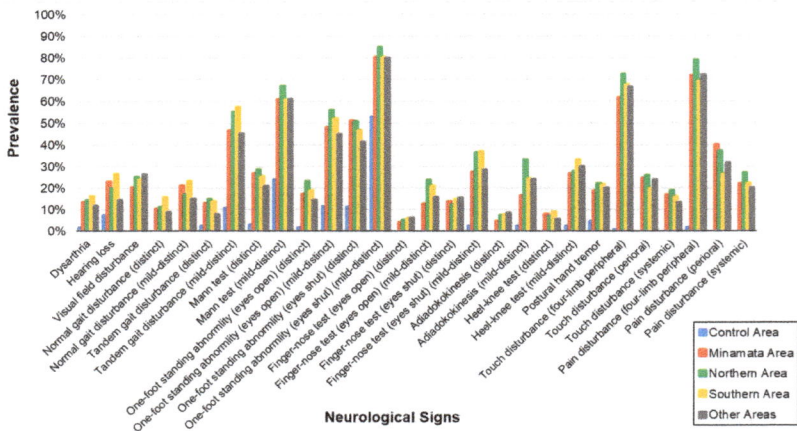

Figure 4. Prevalence of neurological signs in each area. Prevalence was higher in the four exposed groups than that in the control area. The patterns of signs were similar among the four groups.

Table 8. Correlation of prevalence of neurological findings among each area.

Correlation/(*p*-Value)	Control Area	Minamata Area	Northern Area	Southern Area
Minamata Area (*p*-value)	0.6709 (0.0001)	1.0000		
Northern Area (*p*-value)	0.6327 (0.0003)	0.9792 (0.0000)	1.0000	
Southern Area (*p*-value)	0.6802 (0.0001)	0.9716 (0.0000)	0.9823 (0.0000)	1.0000
Other Areas (*p*-value)	0.6518 (0.0002)	0.9778 (0.0000)	0.9864 (0.0000)	0.9771 (0.0000)

3.4. Scores of Signs and Symptoms

Scores of signs and symptoms were shown in Table 9. All scores were significantly higher and also similar in the four exposed groups. After recalculating the data in DA, NDA, and BA1968, even NDA and BA1968, scores were extremely high in the exposed groups (Table 10).

Table 9. Score of signs and symptoms in each area (*n* = 1115).

	Control Area	Minamata Area	Northern Area	Southern Area	Other Areas
	(*n* = 142)	(*n* = 259)	(*n* = 279)	(*n* = 246)	(*n* = 189)
Symptom score (always)					
Mean $\pm$ SD	0.2 $\pm$ 0.5	5.1 $\pm$ 4.9	6.1 $\pm$ 5.0	6.9 $\pm$ 5.4	5.3 $\pm$ 4.9
Range (min–max)	0–3	0–22	0–23	0–22	0–23
Symptom score (always and sometimes)					
Mean $\pm$ SD	3.2 $\pm$ 3.0	21.2 $\pm$ 9.2	23.3 $\pm$ 9.0	24.7 $\pm$ 9.2	21.9 $\pm$ 9.4
Range (min–max)	0–15	2–45	1–46	3–45	2–46
Cranial nerve score					
Mean $\pm$ SD	0.2 $\pm$ 0.6	1.1 $\pm$ 1.7	1.1 $\pm$ 1.5	1.4 $\pm$ 1.8	1 $\pm$ 1.5
Range (min–max)	0–2	0–6	0–6	0–6	0–6
Upper, lower ataxia and tremor score					
Mean $\pm$ SD	0.1 $\pm$ 0.4	1.0 $\pm$ 1.2	1.3 $\pm$ 1.5	1.3 $\pm$ 1.5	1.1 $\pm$ 1.4
Range (min–max)	0–3	0–5	0–5	0–5	0–5
Truncal ataxia score					
Mean $\pm$ SD	0.9 $\pm$ 1.0	2.5 $\pm$ 1.7	2.7 $\pm$ 1.6	2.7 $\pm$ 1.6	2.4 $\pm$ 1.6
Range (min–max)	0–3	0–5	0–5	0–5	0–5
Sensory score					
Mean $\pm$ SD	0.0 $\pm$ 0.2	2.4 $\pm$ 1.5	2.6 $\pm$ 1.5	2.2 $\pm$ 1.3	2.3 $\pm$ 1.2
Range (min–max)	0–2	0–6	0–6	0–6	0–5
Total neurological score					
Mean $\pm$ SD	1.2 $\pm$ 1.4	6.9 $\pm$ 4.5	7.7 $\pm$ 4.2	7.6 $\pm$ 4.3	6.8 $\pm$ 4.0
Range (min–max)	0–8	0–20	0–21	0–21	0–19

Table 10. Score of signs and symptoms in DA, NDA, and BA1968 (*n* = 1115).

	Control Area	DA	NDA	BA1968
Age (Mean $\pm$ SD)	62.0 $\pm$ 10.5	63.0 $\pm$ 11.3	63.6 $\pm$ 10.0	37.4 $\pm$ 2.3
(*n*)	(142)	(786)	(158)	(30)
Symptom score (always)				
Mean $\pm$ SD	0.2 $\pm$ 0.5	5.9 $\pm$ 5.2	6.1 $\pm$ 4.9	4.0 $\pm$ 5.0
Range (min–max)	0–3	0–23	0–23	0–19
Symptom score (always and sometimes)				
Mean $\pm$ SD	3.2 $\pm$ 3.0	22.9 $\pm$ 9.2	23.1 $\pm$ 9.2	19.7 $\pm$ 9.9
Range (min–max)	0–15	2–46	1–46	2–42
Cranial nerve score				
Mean $\pm$ SD	0.2 $\pm$ 0.6	1.2 $\pm$ 1.7	1.1 $\pm$ 1.6	0.4 $\pm$ 0.8
Range (min–max)	0–2	0–6	0–6	0–2
Upper, lower ataxia and tremor score				
Mean $\pm$ SD	0.1 $\pm$ 0.4	1.2 $\pm$ 1.4	1.3 $\pm$ 1.5	1.0 $\pm$ 1.2
Range (min–max)	0–3	0–5	0–5	0–4
Truncal ataxia score				
Mean $\pm$ SD	0.9 $\pm$ 1.0	2.6 $\pm$ 1.6	2.7 $\pm$ 1.7	1.7 $\pm$ 1.3
Range (min–max)	0–3	0–5	0–5	0–4
Sensory score				
Mean $\pm$ SD	0.0 $\pm$ 0.2	2.4 $\pm$ 1.4	2.4 $\pm$ 1.4	2.0 $\pm$ 1.7
Range (min–max)	0–2	0–6	0–6	0–6
Total neurological score				
Mean $\pm$ SD	1.2 $\pm$ 1.4	7.3 $\pm$ 4.3	7.5 $\pm$ 4.2	5.1 $\pm$ 3.8
Range (min–max)	0–8	0–21	0–20	0–13

Relations between scores and frequency of fish ingestion are shown in Table 11 and Figure 5. The symptom (always) score was higher when the frequency of fish ingestion was more. Although scores in the "<1/week" group were higher than in the "1/day" and "≥1/week" groups, the causes were unknown, so we merged the scores in "≥1/week" and "<1/week" into "<1/day" in Figure 5.

The symptom (always and sometimes) score was higher when frequency of fish ingestion was more. As with the symptom (always) score, scores in the "<1/week" group were higher than those in the "≥1/week" group, the causes were unknown, so we merged the scores in "≥1/week" and "<1/week" into "<1/day" in Figure 5.

The total neurological score was higher when frequency of fish ingestion was more (Table 11, Figure 5).

Table 11. Frequency of fish ingestion and scores (*n* = 973).

Frequency	*n*		Age	Symptom (Always) Score	Symptom (Always & Sometimes) Score	Total Neurological Score
3/day	459	Mean ± SD	62.9 ± 11.6	6.7 ± 5.5	24.4 ± 9.3	7.9 ± 4.3
		Min–Max	33–90	0–23	3–46	0–21
2/day	271	Mean ± SD	62.2 ± 11.5	5.9 ± 4.7	23.0 ± 8.8	7.1 ± 4.3
		Min–Max	36–92	0–21	1–44	0–21
1/day	146	Mean ± SD	61.9 ± 12.7	4.8 ± 4.7	20.8 ± 8.7	7.0 ± 4.3
		Min–Max	34–89	0–21	2–44	0–20
≥1/week	79	Mean ± SD	61.0 ± 11.3	3.4 ± 3.7	17.0 ± 8.4	5.7 ± 3.7
		Min–Max	35–86	0–15	2–36	0–16
<1/week	18	Mean ± SD	58.9 ± 12.3	5.1 ± 4.6	20.7 ± 10	4.4 ± 3.3
		Min–Max	39–81	0–16	4–38	1–11
<1/day (≥1/week & <1/week)	97	Mean ± SD	60.6 ± 11.4	3.7 ± 3.9	17.7 ± 8.8	5.5 ± 3.6
		Min–Max	35–86	0–16	2–38	0–16
Total	973	Mean ± SD	62.3 ± 11.7	5.9 ± 5.1	22.8 ± 9.3	7.3 ± 4.3
		Min–Max	33–92	0–23	1–46	0–21

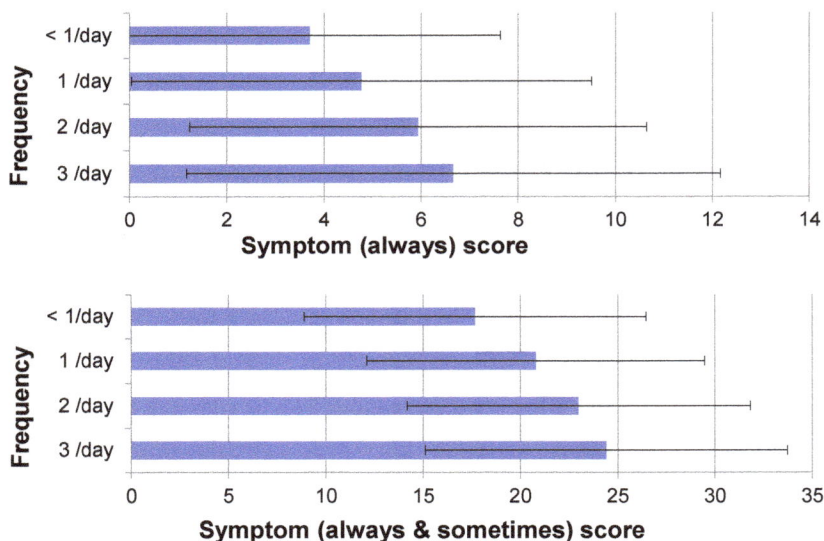

Figure 5. *Cont.*

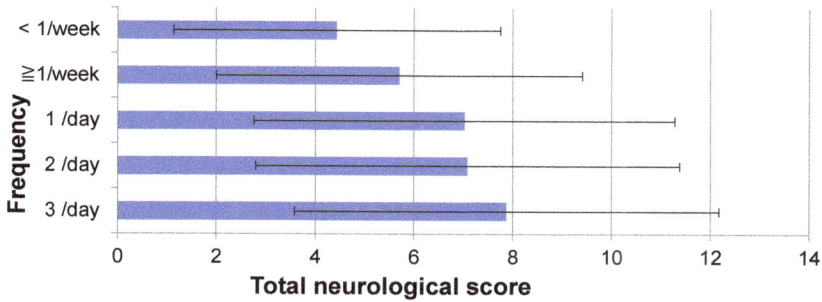

Figure 5. Scores and frequency of fish ingestion (*n* = 973). Scores increased as the frequency of fish ingestion increased.

In each of the four neurological scores (cranial nerve score, upper, lower ataxia and tremor score, truncal ataxia score, and sensory score), the higher the frequency of fish ingestion, the more the score increased. But the range of scores in each frequency group was greater between the zero to maximum score (Figure 6).

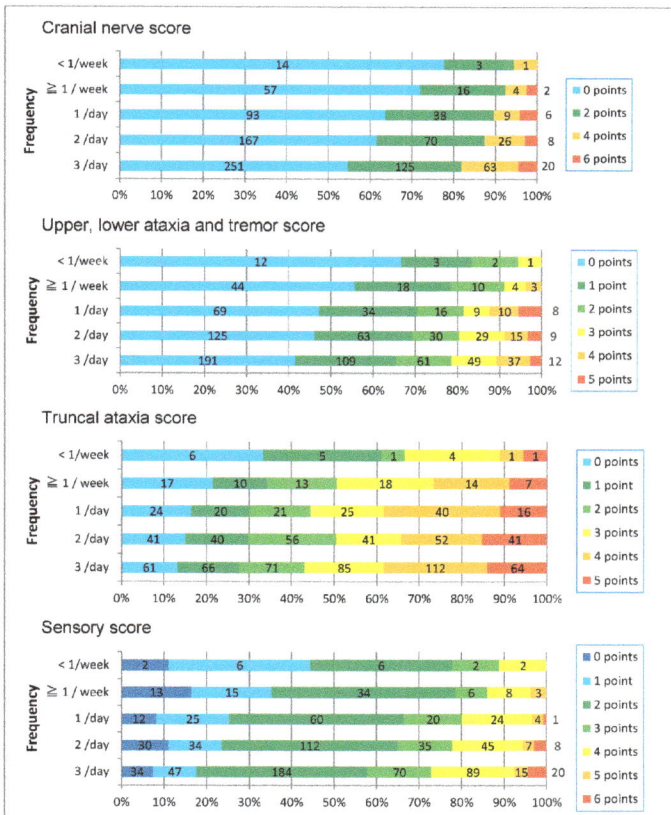

Figure 6. Each score increased as the frequency of fish ingestion increased.

3.5. Onset of Symptoms

3.5.1. Onset of Symptoms in Each Group

Table 12 shows onset year of symptoms in each group. Average year of the first symptom is 1979.0 ± 14.8 in the Minamata Area. That is more than 10 years after the Chisso Factory stopped releasing polluted waste-water in May 1968. Sixty-five percent (628/996) in the exposed groups displayed the first symptom later than 1968. The onset of the first symptom, muscle cramps, four-limb numbness, stumbling tendency, difficulty in fine finger task, and limited peripheral vision were slightly later in subjects in the Minamata Area, but almost the same among four exposed groups. In the Minamata Area, more residents had already been examined for methylmercury poisoning than in other groups, which might be related to the reason why the average year of onset was later than other places.

Muscle cramps occurred 4.8 ± 8.9 years after the first symptom. Four-limb numbness (8.7 ± 11.8 years), difficulty in fine finger task (13.2 ± 12.7 years), stumbling tendency (14.3 ± 13.5 years), and limited peripheral vision (17.3 ± 13.3 years) were followed. Figure 7 shows the total number of subjects with each symptom.

Table 12. Average onset year of symptoms and interval between the first symptom and the onset of each following symptom in each area.

	Minamata Area	Northern Area	Southern Area	Other Areas	Total of Polluted Areas
First symptom	1979.0 ± 14.8	1974.9 ± 16.1	1973.6 ± 14.2	1974.6 ± 13.7	1975.6 ± 15
case/N (%)	256/259 (98.8)	275/279 (98.6)	245/246 (99.6)	189/189 (100)	965/973 (99.2)
>1968/case (%)	185/256 (72.3)	175/275 (63.6)	143/245 (58.4)	124/189 (65.6)	627/965 (65.0)
Muscle cramps	1982.2 ± 15.1	1979.5 ± 15.6	1978.1 ± 15.1	1978.4 ± 14.1	1979.6 ± 15.1
case/N (%)	235/259 (90.7)	255/279 (91.4)	227/246 (92.3)	182/189 (96.3)	899/973 (92.4)
>1968/case (%)	185/235 (78.7)	192/255 (75.3)	157/227 (69.2)	139/182 (76.4)	673/899 (74.9)
First symptom-Muscle cramps	4.4 ± 8.8	5.5 ± 10.4	4.9 ± 8.6	4.2 ± 7.2	4.8 ± 8.9
Four limb numbness	1987.8 ± 14.2	1982.2 ± 15.9	1982.3 ± 14.5	1983.9 ± 13.9	1984 ± 14.9
case/N (%)	237/259 (91.5)	254/279 (91.0)	235/246 (95.5)	172/189 (91.0)	898/973 (92.3)
>1968/case (%)	205/237 (86.5)	205/254 (80.7)	189/235 (80.4)	151/172 (87.8)	750/898 (83.5)
First symptom-Four limb numbness	9.1 ± 12.5	8.1 ± 12.0	8.5 ± 11.2	9.4 ± 11.4	8.7 ± 11.8
Stumbling tendency	1990.7 ± 13.1	1989.2 ± 13.7	1987.9 ± 14.3	1986.8 ± 13.4	1988.8 ± 13.7
case/N (%)	188/259 (72.6)	216/279 (77.4)	196/246 (79.7)	144/189 (76.2)	744/973 (76.5)
>1968/case (%)	176/188 (93.6)	196/216 (90.7)	175/196 (89.3)	133/144 (92.4)	680/744 (91.4)
First symptom-Stumbling tendency	12.7 ± 13.2	15.4 ± 13.2	14.9 ± 14.0	14.1 ± 13.7	14.3 ± 13.5
Difficulty in fine finger tasks	1989.5 ± 13.6	1986.9 ± 14.4	1985.8 ± 13.5	1983.1 ± 13.8	1986.5 ± 14.0
case/N (%)	144/259 (55.6)	182/279 (65.2)	178/246 (72.4)	114/189 (60.3)	618/973 (63.5)
>1968/case (%)	135/144 (93.8)	161/182 (88.5)	157/178 (88.2)	95/114 (83.3)	548/618 (88.7)
First symptom-Difficulty in fine finger tasks	12.3 ± 12.7	14.4 ± 13.1	13.4 ± 12.9	12.1 ± 12.0	13.2 ± 12.7
Limited peripheral vision	1992.2 ± 13.8	1988.8 ± 14.6	1991.9 ± 10.9	1988.9 ± 12.7	1990.5 ± 13.2
case/N (%)	134/259 (51.7)	179/279 (64.2)	168/246 (68.3)	110/189 (58.2)	591/973 (60.7)
>1968/case (%)	126/134 (94.0)	160/179 (89.4)	165/168 (98.2)	103/110 (93.6)	554/591 (93.7)
First symptom-Limited peripheral vision	16.5 ± 13.6	16.9 ± 12.9	18.7 ± 13.1	16.8 ± 13.7	17.3 ± 13.3

The appearance of symptoms was almost the same among the four exposed groups. Figure 8 shows the cumulative rate of onset year of the first symptom in each exposed group.

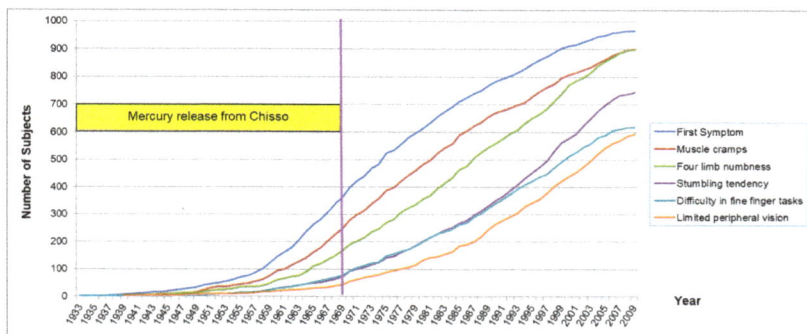

Figure 7. Onset year of each neurological symptom (actual number). Sixty-five percent of the subjects in the exposed area experienced their first symptoms after 1968, when polluted waste water from the Chisso Company's factory was halted. Symptoms have even appeared in recent years.

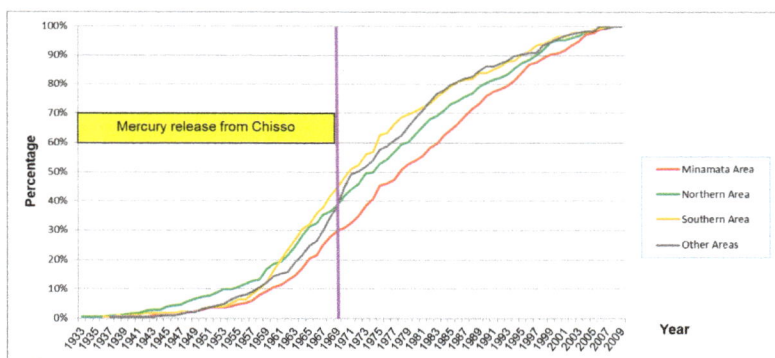

Figure 8. Onset year of the first symptom in each exposed group (percentage). Cumulative curves of the onset were almost the same among the four exposed groups.

3.5.2. Relation between the Onset of Symptoms and Frequency of Fish Ingestion

The frequency of fish ingestion was closely related to the onset year of other symptoms (Table 13, Figure 9). Figure 10 shows duration between the first symptom and each other symptom. The duration between the first symptom and four-limb numbness was related to the frequency of ingestion, but there were almost no relations in other symptoms.

Table 13. Average onset year of symptoms and interval from the first symptom in each category of fish ingestion ($n = 965$).

	3/day	2/day	1/day	$\geq$1/week	<1/week
	($n = 455$)	($n = 269$)	($n = 144$)	($n = 79$)	($n = 18$)
First symptom	1974.2 ± 15.1	1975.6 ± 14.6	1978.2 ± 14.8	1977.7 ± 15	1982.5 ± 15.2
>1968/case (%)	277/455 (60.9)	174/269 (64.7)	105/144 (72.9)	56/79 (70.9)	15/18 (83.3)
Muscle cramps	1978 ± 15.1	1980.6 ± 14.9	1981.7 ± 14.5	1980.9 ± 15.5	1986.4 ± 17
case/N (=965) (%)	428/455 (94.1)	255/269 (94.8)	133/144 (92.4)	69/79 (87.3)	14/18 (77.8)
>1968/case (%)	308/428 (72.0)	194/255 (76.1)	108/133 (81.2)	52/69 (75.4)	11/14 (78.6)
First symptom-Muscle cramps	4.6 ± 8.2	5.4 ± 10.4	4.9 ± 9.3	3.9 ± 6.3	5.7 ± 10.5

Table 13. *Cont.*

	3/day	2/day	1/day	≥1/week	<1/week
	(*n* = 455)	(*n* = 269)	(*n* = 144)	(*n* = 79)	(*n* = 18)
Four limb numbness	1981.3 ± 15.5	1984.6 ± 13.8	1987.5 ± 14.6	1990.2 ± 12.6	1991.8 ± 11.7
case/N (=965) (%)	430/455 (94.5)	251/269 (93.3)	129/144 (89.6)	73/79 (92.4)	15/18 (83.3)
>1968/case (%)	339/430 (78.8)	216/251 (86.1)	111/129 (86.0)	69/73 (94.5)	15/15 (100)
First symptom-Four limb numbness	7.6 ± 11.1	8.9 ± 11.6	9.6 ± 12.4	13.1 ± 14.4	9.5 ± 13.1
Stumbling tendency	1987.2 ± 13.9	1988.9 ± 14.1	1990.6 ± 12.5	1994.5 ± 11.5	1993.8 ± 10.5
case/N (=965) (%)	368/455 (80.9)	206/269 (76.6)	112/144 (77.8)	48/79 (60.8)	10/18 (55.6)
>1968/case (%)	332/368 (90.2)	185/206 (89.8)	105/112 (93.8)	48/48 (100)	10/10 (100)
First symptom-Stumbling tendency	14.3 ± 13.0	14.2 ± 13.8	13 ± 13.2	18.9 ± 16.2	14.5 ± 15.2
Difficulty in fine finger tasks	1985.4 ± 13.6	1986.3 ± 14.4	1989.1 ± 14.2	1989.1 ± 14.4	1996.9 ± 8.4
case/N (=965) (%)	320/455 (70.3)	168/269 (62.5)	82/144 (56.9)	40/79 (50.6)	8/18 (44.4)
>1968/case (%)	284/320 (88.8)	147/168 (87.5)	73/82 (89.0)	36/40 (90.0)	8/8 (100)
First symptom-Difficulty in fine finger tasks	13.1 ± 12.3	13 ± 12.7	13.9 ± 13.9	13.1 ± 13.7	12.5 ± 16.7
Limited peripheral vision	1989.7 ± 12.8	1990.8 ± 13.3	1991.8 ± 12.9	1991.3 ± 16.1	1993.8 ± 13.8
case/N (=965) (%)	302/455 (66.4)	165/269 (61.3)	80/144 (55.6)	36/79 (45.6)	8/18 (44.4)
>1968/case (%)	281/302 (93.0)	155/165 (93.9)	77/80 (96.3)	33/36 (91.7)	8/8 (100)
First symptom-Limited peripheral vision	17.5 ± 12.9	17.3 ± 13.7	16.3 ± 13.4	19.1 ± 14.3	13.1 ± 16.3

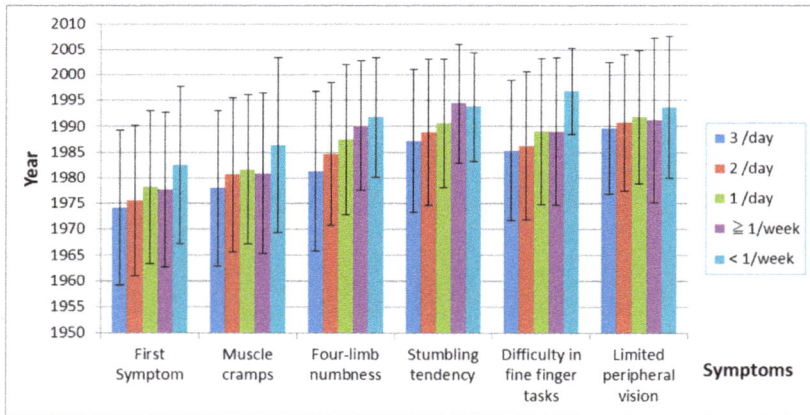

Figure 9. Onset of symptoms and frequency of fish ingestion. The time of onset of subsequent symptoms increased as the frequency of fish ingestion decreased.

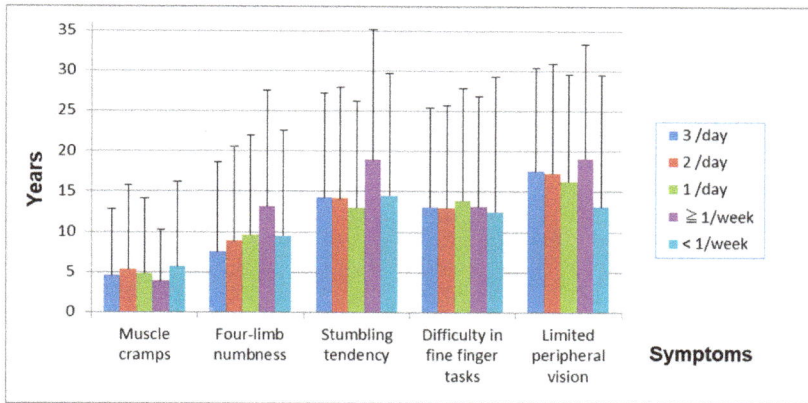

Figure 10. Years of each symptom after the first symptom. There was almost no relation between fish ingestion and duration from the first symptom to each following symptom, except for four-limb numbness.

3.5.3. Relations between Onset of Symptom and Scores of Signs and Symptoms

The onset of symptoms was related to scores. When the onset year was earlier, the score became greater (Table 14).

Table 14. Score of signs and onset of symptoms in each area.

	First Symptom	Muscle Cramps	Four Limb Numbness	Stumbling Tendency	Difficulty in Fine Finger Tasks	Limited Peripheral Vision
Symptom score (always)						
R^2	0.0496	0.0329	0.0789	0.0453	0.0383	0.0567
p-value	(0.000)	(0.000)	(0.000)	(0.000)	(0.000)	(0.000)
Symptom score (always and sometimes)						
R^2	0.0845	0.0612	0.1134	0.0641	0.0443	0.0716
p-value	(0.000)	(0.000)	(0.000)	(0.000)	(0.000)	(0.000)
Cranial nerve score						
R^2	0.0193	0.0127	0.0222	0.0098	0.0105	0.0062
p-value	(0.000)	(0.000)	(0.000)	(0.004)	(0.006)	(0.031)
Upper, lower ataxia and tremor score						
R^2	0.0298	0.0173	0.0438	0.0237	0.0121	0.0141
p-value	(0.000)	(0.000)	(0.000)	(0.000)	(0.004)	(0.002)
Truncal ataxia score						
R^2	0.0143	0.0046	0.0291	0.0165	0.0115	0.008
p-value	(0.023)	(0.000)	(0.000)	(0.004)	(0.017)	(0.000)
Sensory score						
R^2	0.0219	0.0163	0.0309	0.0272	0.022	0.0058
p-value	(0.000)	(0.000)	(0.000)	(0.000)	(0.000)	(0.035)
Total neurological score						
R^2	0.0428	0.0248	0.0633	0.0385	0.0289	0.0189
p-value	(0.000)	(0.000)	(0.000)	(0.000)	(0.000)	(0.000)

3.6. Health Effect in Younger Generation

3.6.1. Characteristics of Younger Subjects in the Exposed and the Control Area

The demographic characteristics of subjects in younger subjects are shown in Table 15. Many of the younger subjects who were born later than 1968 (BA1968) in the exposed areas had also lived in fishermen families (30.0%) and had eaten more fish (3 times/day in 50.0%) than the elderly exposed subjects. While the prevalence of subjects whose parents were fishermen and the prevalence of subjects who had a family history of Minamata disease were almost the same between BA1968 (n = 30, in Table 15) and Designated Area (n = 84, in Table 15), there was no significant difference in complications between BA1968 and the Control Area (n = 88, in Table 15).

Table 15. Demographic characteristics of subjects in younger subjects (n = 202).

	Control Area	BA1968	Designated Area	Total
	(n = 88)	(n = 30)	(n = 84)	(n = 202)
Sex, n (%)				
Male	40 (45.5)	21 (70.0)	44 (52.4)	105 (52.0)
Female	48 (54.6)	9 (30.0)	40 (47.6)	97 (48.0)
Age				
Mean ± SD	37.5 ± 6.0	37.4 ± 2.3	44.8 ± 2.3	40.5 ± 5.6
Range (min–max)	30–48	33–40	40–48	30–48
Smoking, n (%)				
Non-smoker	56 (65.1)	19 (63.3)	49 (58.3)	124 (62.0)
Smoker	30 (34.9)	11 (36.7)	35 (41.7)	76 (38.0)
Alcohol drinking, n (%)				
Non-drinker	41 (47.7)	11 (36.7)	30 (35.7)	82 (41.0)
Drinker	45 (52.3)	19 (63.3)	54 (64.3)	118 (59.0)
Frequency of fish intake, n (%)				
Three times a day	3 (3.4)	15 (50.0)	30 (35.7)	48 (23.8)
Twice a day	1 (1.1)	6 (20.0)	24 (28.6)	31 (15.4)
Once a day	9 (10.2)	5 (16.7)	21 (25.0)	35 (17.3)
More than once a week	53 (60.2)	2 (6.7)	8 (9.5)	63 (31.2)
Less than once a week	20 (22.7)	2 (6.7)	1 (1.2)	23 (11.4)
Occupation, n (%)				
Fishermen (subject)	0 (0.0)	1 (3.3)	5 (6.0)	6 (3.0)
Fishermen (subject's parent)	1 (1.2)	9 (30.0)	23 (27.4)	33 (16.6)
Complications, n (%)				
Hypertension	1 (1.1)	0 (0.0)	8 (9.5)	9 (4.5)
Renal diseases	0 (0.0)	0 (0.0)	3 (3.6)	3 (1.5)
Liver diseases	1 (1.1)	0 (0.0)	2 (2.4)	3 (1.5)
Respiratory diseases	8 (9.1)	1 (3.3)	6 (7.1)	15 (7.4)
Diabetes Mellitus	0 (0.0)	1 (3.3)	2 (2.4)	3 (1.5)
Orthopedic diseases	3 (3.4)	3 (10)	14 (16.7)	20 (9.9)
Malignant diseases	0 (0.0)	0 (0.0)	1 (1.2)	1 (0.5)
History of application for inamata disease, n (%)	0 (0.0)	1 (3.3)	13 (15.5)	14 (6.9)
Family history of Minamata disease, n (%)	0 (0.0)	29 (96.7)	76 (90.5)	105 (52.0)
Have witnessed abnormal animal behavior, n (%)	No Data	9 (30.0)	27 (32.1)	36 (31.6)

3.6.2. Symptoms and Neurological Signs in the Younger Generation

Symptoms and neurological signs in younger age (BA1968) were much more prevalent than in the Control Area, and the pattern of positive findings resembled those in the Designated Area (Figures 11–13). The score of signs and symptoms showed similar results (Table 16).

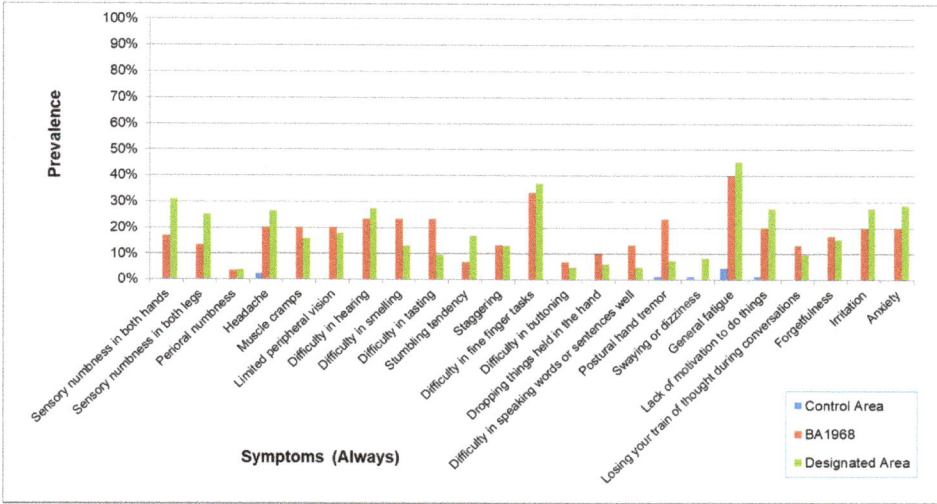

Figure 11. Prevalence of symptoms (Always) in the younger generation. Prevalence in BA1968 and the Designated Area were higher than in the Control Area. The symptomatic patterns in BA1968 and the Designated Area were similar.

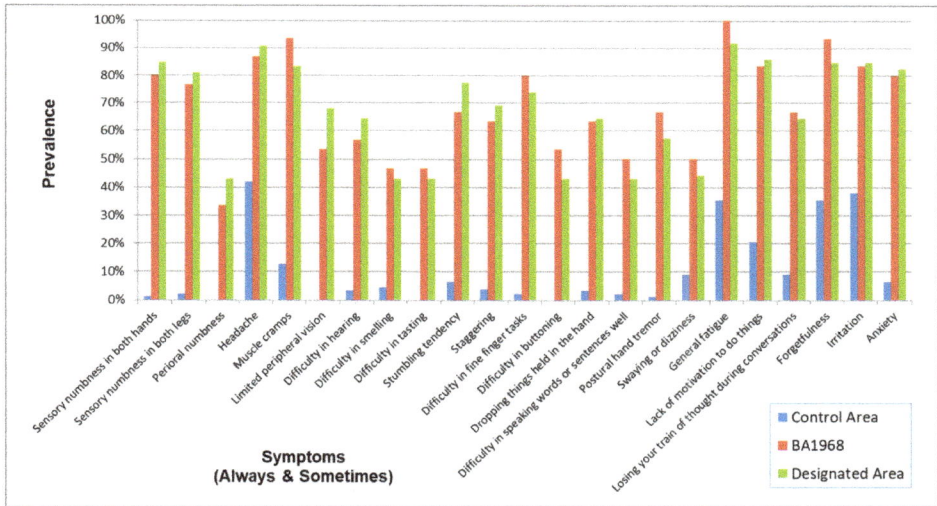

Figure 12. Prevalence of symptoms (Always and Sometimes) in the younger generation. Prevalence in BA1968 and the Designated Area were higher than in the Control Area. The symptomatic patterns in BA1968 and the Designated Area were similar.

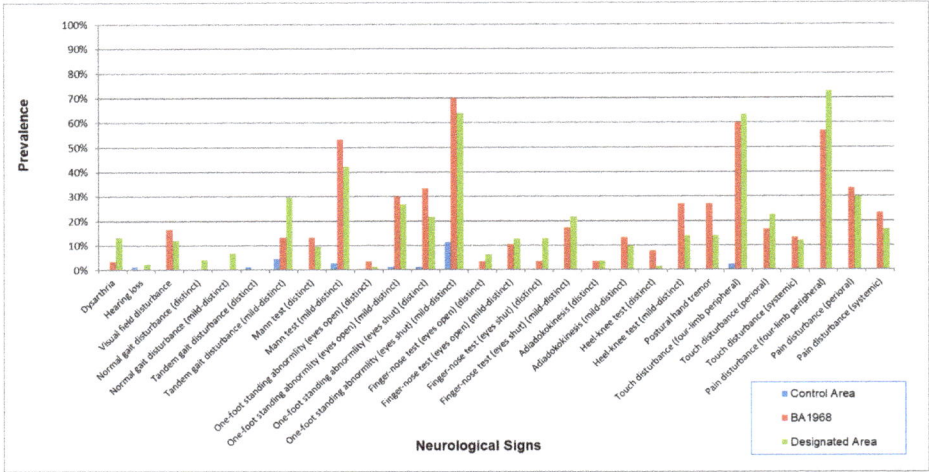

Figure 13. Prevalence of neurological signs in the younger generation. Prevalence in BA1968 and the Designated Area were higher than in the Control Area. The patterns of signs in BA1968 and the Designated Area were similar.

Table 16. Score of signs and symptoms in younger subjects.

	Control Area	BA1968	Designated Area
Age (Mean ± SD)	37.5 ± 10.0	37.4 ± 10.0	44.8 ± 10.0
(*n*)	**(88)**	**(30)**	**(84)**
Symptom score (always)			
Mean ± SD	0.1 ± 0.4	4.0 ± 5.0	4.2 ± 4.3
Range (min–max)	0–3	0–19	0–19
Symptom score (always and sometimes)			
Mean ± SD	2.5 ± 2.9	19.7 ± 9.9	19.8 ± 9.1
Range (min–max)	0–16	2–42	2–42
Cranial nerve score			
Mean ± SD	0.0 ± 0.2	0.4 ± 0.8	0.5 ± 1.1
Range (min–max)	0–2	0–2	0–4
Upper, lower ataxia and tremor score			
Mean ± SD	0.0 ± 0.0	1.0 ± 1.2	0.6 ± 1.0
Range (min–max)	0–0	0–4	0–4
Truncal ataxia score			
Mean ± SD	0.2 ± 0.5	1.7 ± 1.3	1.7 ± 1.5
Range (min–max)	0–3	0–4	0–5
Sensory score			
Mean ± SD	0.0 ± 0.1	2.0 ± 1.7	2.2 ± 1.3
Range (min–max)	0–1	0–6	0–5
Total neurological score			
Mean ± SD	0.2 ± 0.7	5.1 ± 3.8	5.0 ± 3.3
Range (min–max)	0–5	0–13	0–16

3.6.3. Onset of Symptoms of Designated, Non-Designated Area and Subjects Born after 1968

Figure 14 shows the onset of the first symptom in the designated area (DA), non-designated area (NDA) and subjects born after the end of 1968 (BA1968). The classifications for these groups were the same as those in Table 10 (not in Table 15). The cumulative curves were almost the same between subjects who had lived in the designated and non-designated area. 82.8% (24/30) of the subjects who were born after 1968 had developed their first symptom by 1987.

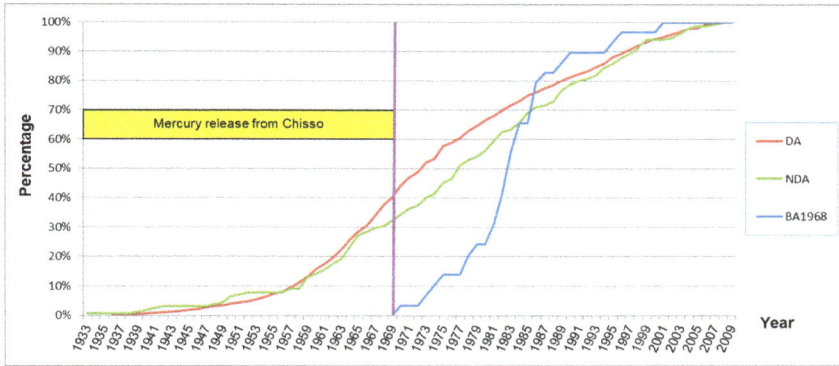

Figure 14. Onset year of the first symptom of subjects with and without residential history in the designated area (percentage).

3.7. Signs and Symptoms in Subjects in Whom Sensory Disturbance had not been Detected during Their Physical Examination

The demographic characteristics of subjects with and without sensory disturbance are shown in Table 17. Overall, the prevalence of symptoms (Always) and symptoms (Always and Sometimes) without sensory disturbance were a little lower than the subjects with some sort of sensory disturbance (Figures 15 and 16). However, most of their symptoms were more prevalent than the control group. Furthermore, the prevalence pattern of neurological signs other than sensory disturbance in subjects without sensory disturbance was similar to that in subjects with some sort of sensory disturbance (Figure 17).

Table 17. Demographic characteristics of subjects in each group (*n* = 1115).

	Control Area	Sensory Disturbance (−)	Sensory Disturbance (+)
	(*n* = 142)	(*n* = 91)	(*n* = 882)
Sex, *n* (%)			
Male	56 (39.4)	64 (70.3)	428 (48.5)
Female	86 (60.6)	27 (29.7)	454 (51.5)
Age			
Mean ± SD	62.0 ± 10.5	59.7 ± 13.3	62.6 ± 11.6
Range (min–max)	36–86	33–87	33–92
Residential history in designated area (DA) more than 1 year, *n* (%)			
In DA > 1 year (DA)	3 (2.1)	69 (75.8)	717 (81.3)
Not in DA > 1 year (NDA)	139 (97.9)	13 (14.3)	144 (16.3)
Born after 1968 (BA1968)	0 (0.0)	9 (5.0)	21 (2.4)

Table 17. *Cont.*

	Control Area	Sensory Disturbance (−)	Sensory Disturbance (+)
	(*n* = 142)	(*n* = 91)	(*n* = 882)
Smoking, *n* (%)			
Non-smoker	109 (77.3)	66 (72.5)	676 (81.0)
Smoker	32 (22.7)	25 (27.5)	206 (19.0)
Alcohol drinking, *n* (%)			
Non-drinker	72 (51.1)	43 (47.3)	468 (53.1)
Drinker	69 (48.9)	48 (52.7)	414 (47.0)
Frequency of fish intake, *n* (%)			
Three times a day	6 (4.4)	34 (37.4)	425 (48.2)
Twice a day	7 (5.1)	30 (33.0)	241 (27.3)
Once a day	27 (19.9)	12 (13.2)	134 (15.2)
More than once a week	63 (46.3)	13 (14.3)	66 (7.5)
Less than once a week	33 (24.3)	2 (2.2)	16 (1.8)
Occupation, *n* (%)			
Fishermen (subject)	0 (0.0)	13 (14.3)	118 (13.4)
Fishermen (subject's parent)	2 (1.6)	28 (30.8)	291 (33.0)
Complications, *n* (%)			
Hypertension	40 (28.2)	28 (30.8)	320 (36.3)
Renal diseases	3 (2.1)	7 (7.7)	48 (5.4)
Liver diseases	6 (4.2)	10 (11.0)	61 (6.9)
Respiratory diseases	12 (8.5)	4 (4.4)	37 (4.2)
Diabetes Mellitus	3 (2.1)	9 (9.9)	84 (9.5)
Orthopedic diseases	13 (9.2)	22 (24.2)	224 (25.4)
Malignant diseases	7 (4.9)	3 (3.3)	48 (5.4)
History of application for Minamata disease, *n* (%)	0 (0.0)	3 (3.3)	109 (12.4)
Family history of Minamata disease, *n* (%)	0 (0.0)	48 (52.7)	498 (56.5)
Have witnessed abnormal animal behavior, *n* (%)	No Data	27 (29.7)	360 (40.8)

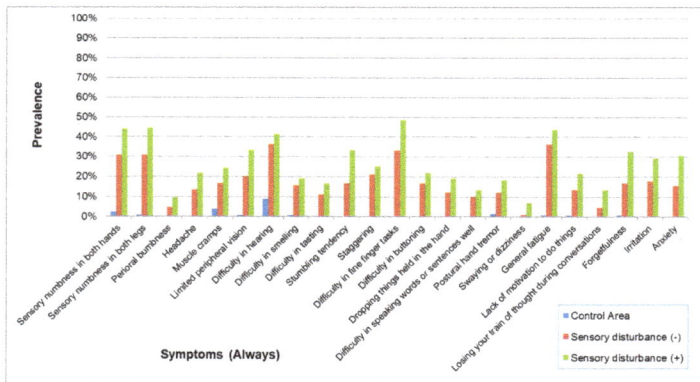

Figure 15. Prevalence of symptoms (Always) in subjects with and without sensory disturbance. The prevalence of symptoms was a little lower in subjects without sensory disturbance than in subjects with sensory disturbance but was apparently higher than the Control Area. The prevalence patterns were similar in exposed subjects with and without sensory disturbance.

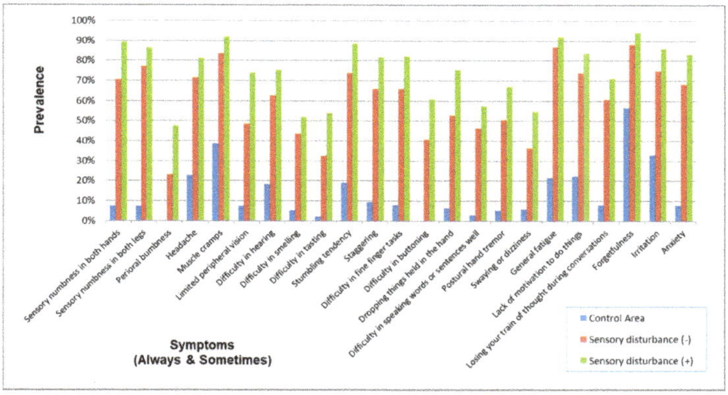

Figure 16. Prevalence of symptoms (Always and Sometimes) in subjects with and without sensory disturbance. The prevalence of symptoms was a little lower in subjects without sensory disturbance than in subjects with sensory disturbance but was apparently higher than the Control Area. The prevalence patterns were similar in exposed subjects with and without sensory disturbance.

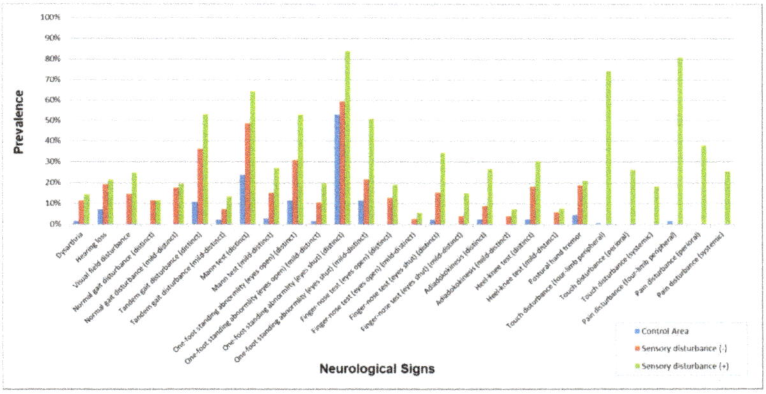

Figure 17. Prevalence of neurological signs in subjects with and without sensory disturbance. Except for the prevalence of sensory disturbance, that of other symptoms was lower in subjects without sensory disturbance than in subjects with sensory disturbance but was generally higher than the Control Area. The prevalence patterns were similar in exposed subjects with and without sensory disturbance except for the prevalence of sensory disturbance.

The scores of signs and symptoms in subjects without sensory disturbance were lower than those in subjects in the exposed area and higher than those in the control area (Table 18).

Table 18. Score of signs and symptoms in subjects with and without sensory disturbance (*n* = 1115).

	Control Area	Sensory Disturbance (−)	Sensory Disturbance (+)
Age (Mean ± SD)	62.0 ± 10.5	59.7 ± 13.3	62.6 ± 11.6
(*n*)	(142)	(91)	(882)
Symptom score (always)			
Mean ± SD	0.2 ± 0.5	4.0 ± 4.0	6.1 ± 5.2
Range (min–max)	0–3	0–16	0–23
Symptom score (always and sometimes)			
Mean ± SD	3.2 ± 3.0	18.0 ± 8.7	23.3 ± 9.2
Range (min–max)	0–15	2–38	1–46
Cranial nerve score			
Mean ± SD	0.2 ± 0.6	0.9 ± 1.4	1.2 ± 1.7
Range (min–max)	0–2	0–6	0–6
Upper, lower ataxia and tremor score			
Mean ± SD	0.1 ± 0.4	0.6 ± 1.0	1.2 ± 1.4
Range (min–max)	0–3	0–4	0–5
Truncal ataxia score			
Mean ± SD	0.9 ± 1.0	1.8 ± 1.7	2.7 ± 1.6
Range (min–max)	0–3	0–5	0–5
Sensory score			
Mean ± SD	0.0 ± 0.2	0.0 ± 0.0	2.6 ± 1.2
Range (min–max)	0–2	0–0	1–6
Total neurological score			
Mean ± SD	1.2 ± 1.4	3.4 ± 3.0	7.7 ± 4.2
Range (min–max)	0–8	0–15	1–21

4. Discussion

4.1. Characteristics of the Subjects

The data collected on the occupations and dietary habits of the four methylmercury-exposed groups showed clearly that the majority are connected to the fishing industry and that their diets consist of large amounts of fish and shellfish.

Certified Minamata disease patients and compensated patients had lived in the designated area, which consisted of the whole Minamata Area, a large part of the Southern Area, and a small part of the Northern Area. In these areas, thousands of residents had already been examined for Minamata disease. But in the non-designated area, which included a part of the Northern and Southern Areas, as well as some other smaller areas, residents had not received information about methylmercury poisoning and had had little opportunity to be examined for methylmercury poisoning. These situations were reflected in the higher percentage of subjects (41.6%) who had not lived in the designated area in the Northern Area than the other three exposed groups (Table 1).

To adjust the higher prevalence of diabetes mellitus and orthopedic diseases in the exposed groups, we used logistic regression analysis. After adjustment for age, sex, diabetes mellitus, and orthopedic diseases, the prevalence of most signs and symptoms were extremely higher in the four exposed groups.

This means that neither diabetes mellitus nor orthopedic diseases were the cause of these signs and symptoms. However, on the contrary, methylmercury exposure might increase diabetes mellitus and orthopedic diseases. Shigenaga reported that in infantile and acute adult Minamata disease cases, injury to the pancreatic islet cells occurred [7]. Harada reported that deformity of four limbs (19%) and four-limb pain (36%) were observed in 145 family members of certified Minamata disease patients [8].

4.2. Characteristics of Signs and Symptoms of Chronic Methylmercury Poisonings

In this study, symptoms of methylmercury-intoxicated subjects varied widely. In 1973, Tatetsu et al. reported symptoms in 215 patients whom they diagnosed with Minamata disease and examined precisely using a two-point scale ("yes" or "no"). The symptoms (prevalence) included numbness of hands and feet (82%), dysesthesia of hands and feet (50%), pain in the head, back, lower back, four limbs (79%), fatigue (65%), visual disturbance (55%), limited peripheral vision (31%), difficulty in hearing (60%), difficulty in smelling (28%), difficulty in tasting (22%), stumbling tendency (58%), difficulty in fine finger tasks (37%), difficulty in buttoning (34%), difficulty in speaking (41%), muscle cramps (71%), hand tremor (43%), insomnia (43%), and forgetfulness (75%) [9].

From 1974 to 1979, Fujino used a two-point scale questionnaire to interview adult residents of Katsurajima Island, Izumi City and reported general fatigue (100%), forgetfulness, difficulty in calculation, inability to concentrate (95%), numbness (95%), difficulty in hearing (90%), muscle cramps (90%), insomnia (90%), staggering or stumbling (93%), difficulty in buttoning (88%), hand tremor (71%), limited peripheral vision (78%), difficulty in smelling (68%) in 41 adult residents [10]. The lower the age of the Katsurajima Island residents, the milder the symptoms.

In 1985, Kinjo et al. used a two-point scale questionnaire to interview certified Minamata disease patients in these designated areas. In Kinjo's study, constriction of visual field (19.3%), difficulty in hearing (54.0%), difficulty in speaking (38.2%), difficulty in buttoning (52.8%), stumbling (69.3%), tremor (39.5%), hypoesthesia of the limbs (67.1%), dysesthesia of the limbs (88.4%), hypoesthesia of the mouth (25.4%), forgetfulness (88.4%), fatigue (82.9%), cramps (80.0%) were observed [11]. 24 years later the variety of symptoms was the same in non-certified residents as had been observed earlier in the certified patients.

In 2005, we performed a study of residents who had been exposed to methylmercury. The symptoms in the group without neurological complications (Always and Sometimes) were numbness of hands and feet (89%), limited peripheral vision (66%), difficulty in hearing (61%), stumbling tendency (63%), difficulty in buttoning (54%), muscle cramps (97%), hand tremor (75%), fatigue (88%), and forgetfulness (97%) [6].

The prevalence pattern of symptoms (Always and Sometimes) were similar to questions common to three of the earlier studies—Tatetsu et al., Kinjo et al. and Takaoka (2005), as well as the present Takaoka study. Fujino's study showed much more severe symptoms than the four previously mentioned studies (Figure 18). The similarities in symptom patterns may be due to the fact that the effects of methylmercury poisoning on health have persisted into the twenty-first century. Because there were differences in these studies in the selection of subjects, phrasing of questions, and the scoring methods' choice of answers were not necessary the same, the percentages of each study, shown below, do not necessarily represent the severity of subjects' symptoms.

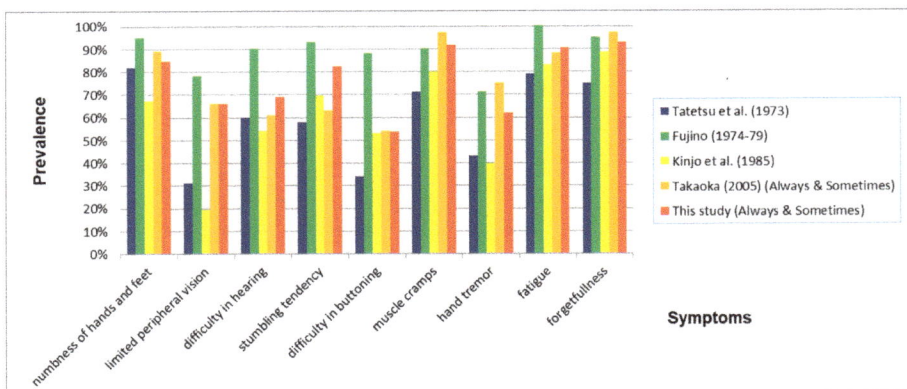

Figure 18. Prevalence of symptoms when comparing nine questions common to all five studies. The prevalence of all characteristic symptoms for Minamata disease was very. In this study, we used the prevalence of "sensory numbness in both hands" instead of "numbness of hands and feet".

Sugiura studied fishermen in Minamata City (1990) and in Izumi City (1988) and pointed out that health problems were present not only in certified patients but also in uncertified patients and other fishermen [12]. In our study, there were no certified patients, but most of them had similar symptoms.

Although activities of daily living (ADL) of these patients decreased from 60 years old, 9.1% still needed assistance in eating, 11.6% in bodily hygiene and 10.6% in using the toilet [11]. The ADL level of most subjects in this study was "Independent" so they were able to come for the examination without serious assistance. The effects of chronic methylmercury cover a wide range, and disabilities remain or progress slowly in many cases.

The symptoms in the questionnaire consist of those both specific and non-specific to methylmercury poisoning. The symptoms whose percentage of the answer (always and sometimes) in the Control Area were considerably lower, compared to those in the exposed groups (e.g., "perioral numbness", "difficulty in tasting", "difficulty in buttoning"), are supposed to more specific symptoms. Tables 2 and 4, Figures 2 and 3 show that prevalence of specific symptoms as well as that of non-specific symptoms became higher through methylmercury exposure.

Increases in specific symptoms from methylmercury poisoning mean that those high percentages of symptoms include the effects of methylmercury. Also increases of non-specific symptoms in exposed people mean that methylmercury also has non-specific health effects.

The prevalence of symptoms was highly correlated among the four exposed groups. By the effects of non-specific symptoms, there were some correlations between the control and the four exposed groups, but the correlations were extremely high among the exposed groups (Tables 3 and 4, Figure 2). These patterns of symptoms were supposed to reflect the characteristics of methylmercury poisonings.

That the correlation between the control and exposed groups in symptoms of "always and sometimes" became higher than that in symptoms of "always" can be explained by effects of the non-specific symptoms for methylmercury poisoning in the control area (Tables 5 and 6, Figure 3).

As with symptoms, positive neurological signs in the four exposed groups were highly more prevalent than the control area (Table 7, Figure 4). Also, the correlation among the four exposed groups was also supposed to reflect the characteristics of methylmercury poisoning (Table 8). Furthermore, the concordance between these findings by 144 doctors with different specialties suggests that the instructions for the examinations were successful.

Many factors may have influence on the final prevalence including possible bias in the selection of subjects, variations in the doctors' examination technique, judgment criteria, the subjects' age, and other health complications. Despite the possible influence of these factors, the high level of consistency in the gathered results show that there was a high degree of precision involved. Also, the collected data suggests that the effects of methylmercury contamination are still present today.

Clinical signs were also as various as studies previously reported. According to Tatesu et al., 269 residents in Minamata district, diagnosed with Minamata disease in 1972, showed sensory disturbance (97.0%), ataxia (93.7%), hearing disturbance (84.0%), weakness (67.3%), speech disturbance (62.5%), and constriction of visual field (59.5%) [9].

Fujino, in his study from 1974 to 1979, reported that those who had lived longer on Katsurajima Island, showed a higher prevalence in all signs of Hunter Russell syndrome: somatosensory disturbance, ataxia, constriction of visual fields, auditory disturbance, and dysarthria. The combination of these symptoms became less as the subjects' ages decreased [10]. In a study from 1975 to 1979, Ninomiya et al. reported that hypoesthesia (60.3%), ataxia (20.7%), impairment of hearing (43.8%), visual change (25.6%), and dysarthria (13.2%) were recognized in 121 residents from the polluted area in Goshonoura [13].

In our study from 2005, subjects in the methylmercury-polluted area had dysarthria (23%), auditory disturbance (33%), visual constriction (28%), positive finger-nose test (50%), heel-shin test (48%), normal gait disturbance (41%), and poor one-foot standing (66%) [6].

Because these studies differed in the selection of subjects, examination method, and judgement criteria, their positive neurological findings were not necessarily the same. However, common abnormalities in many functions including somatosensory, visual, and hearing acuities, coordination of upper and lower extremities were recognized, indicating that methylmercury poisoning still prevails.

4.3. Dose-Response Relationship of Chronic Methylmercury Poisoning

The severity of the subjects' symptoms varied widely and some of them were experienced sporadically, which means that symptoms were persistent, intermittent, or periodical in many cases. The severity of neurological signs ranged from no finding to continual seriousness. The seriousness generally correlated with the volume of fish ingested (Table 11, Figures 5 and 6), but even between subjects with a low intake of fish and those with a high intake the severity of the cases varied from no findings to findings with high severity were observed (Figure 6). This is the first study that dose-response effects were observed in methylmercury poisoning in Japan.

According to pathological studies on Minamata disease, a spongy state, from complete loss of neurons in the cerebral and cerebellar cortex, can occur in the severe cases, whereas in milder cases, "scattered single cell necrosis" in the cortex is supposed to occur [14]. Nowadays, this cellular loss in milder cases has been supposed to be caused by "apoptosis" [15]. This could explain the variability in seriousness of this disease.

A higher percentage of intermittent or periodical signs and symptoms has been supposed be characteristics of mild and moderate chronic methylmercury poisoning. Uchino reported that in 63 out of 77 certified Minamata disease patients (82%), changes in the range of sensory disturbance were observed [16]. The functions of the brain cortex have plasticity and can be affected by other parts of central nervous system, which can lead to fluctuations in signs and symptoms.

4.4. Late Onset of Methylmercury Poisoning

In this study, 65.0% of the subjects in the four exposed areas developed the first symptoms after 1968 (Table 12, Figure 7), when the dumping of polluted waste-water from the Chisso Cooperation stopped. After the appearance of the first symptom, the order of subsequent symptoms appeared gradually in the following order: muscle cramps, four-limb numbness, difficulty in fine finger tasks, stumbling tendency, and limited peripheral vision (Table 12). When we consider that most of the subjects had been exposed to methylmercury in or before 1968, there was a delay in the appearance of symptoms. In 1991, The Central Environmental Pollution Council of the Environment Agency (the predecessor of the present Ministry of Environment) (Japan) reported without evidence, "the period from methylmercury exposure to the onset is ordinarily supposed to be one month, at least within one year" [17]. Although this opinion is held by the Japanese administration even now, the reality is completely different. In the same year 1991, Igata, the chairman of the Council, wrote "roughly speaking, such late onset and late progression are limited to some patients and the peak was reached before 1975" [18], which was 7 years after the Chisso company stopped releasing contaminated wastewater in 1968 and also a time when cases with an acute onset of Minamata disease were rarely found. The true latency period is much longer.

Even in acute cases of maximum methylmercury exposure in Iraq, the mean latent periods ranged from 16 to 38 days [19]. In the experiments with monkeys, a latency period of 6 years was reported [20]. Evans et al. conducted a long-term study on nonhuman primates and demonstrated that the latency period was dose dependent [21,22]. Our data support the thesis that the latency period became longer as the exposure became milder (Table 13, Figure 9). The longest latency from exposure to symptomatic onset has not been determined. According to our data, the first symptom may already have been present before 2008, one year before the survey (Figures 7 and 8), which explains that it is difficult to determine the longest latency period.

4.5. How Far Had the Methylmercury Pollution Spread?

Many of the subjects in the Northern Area (41.6%) had not lived in the designated area, but the pattern of signs and symptoms were almost the same as the Minamata Area (4.6%), Southern Area (7.7%), and Other Areas (5.3%) (Table 2). The scores of signs and symptoms in the non-designated area were almost the same as the designated area (Table 10). These data show that residents in the non-designated area had similar health effects from methylmercury and the spread of Minamata disease was larger than previously thought. At least, within the area where fish and shellfish can be obtained daily, adverse health effects caused by methylmercury may have occurred.

4.6. What Is the Longest Time, After Methylmercury Exposure, That Late Developing Symptoms Can Appear?

Similar patterns of signs and symptoms were observed in BA1968 (Table 15, Figures 11–13) as same as in the elderly subjects. The scores of signs and symptoms in BA1968 were also similar to those of subjects in the designated area, and greater than the Control Area (Table 16). Figure 14 shows the development of methylmercury-related symptoms. The data indicates that the detrimental effects on health from methylmercury poisoning had continued to spread, even after the release of polluted waste-water had been stopped in 1968.

The Central Environmental Pollution Council of the Environment Agency, without any evidence, stated in 1991, "Since 1969, the possibilities of being exposed to levels of methylmercury that can cause Minamata disease no longer exist". Our study shows that this statement in 1991 was incorrect.

It is difficult to determine whether subjects of BA1968 (as displayed in Figure 14) had developed their symptoms due to continued exposure after 1968 or if they were late developing symptoms. What we can say is that it is difficult to determine at which time, after chronic or continuous methylmercury exposure, that a person, not showing health problems, can be judged to be safe from late developing Minamata Disease symptoms.

4.7. Signs and Symptoms in Subjects Whose Sensory Disturbance Was Not Recognized

In Japan, there are a lot of people who have been exposed methylmercury and who have some neurological abnormalities. Four-limb sensory disturbance has been supposed to be the minimum neurological abnormality in Japan [1]. Outside of Japan, epidemiological studies have not seen such extreme neurological signs, but other more mild or latent neurocognitive and behavioral symptoms [23,24].

This study showed that subjects without sensory disturbance had experienced many other subjective symptoms and objective abnormalities similar to those of subjects with sensory disturbance (Table 17, Figures 15–17). This means that there is a range of varieties of symptoms from methylmercury poisoning. Although sensory disturbance is important in methylmercury poisoning, the recognition of health problems caused by methylmercury should not be limited to cases with sensory disturbance.

4.8. Other Problems

There are some limitations in this study. Firstly, this study was a cross-sectional study and participants in this study were limited to subjects who had volunteered to be tested. Therefore, it is obvious that our data may not concur with the data of the general population in these areas. But our data does show the spread of methylmercury poisoning along Shiranui Seashore.

Secondly, the lifestyle and occupations were not the same among the four groups. But the high prevalence of the specific complaints and neurological findings of Minamata disease and the similar patterns of such prevalence of symptoms and neurological signs support our opinion.

5. Conclusions

The effects of methylmercury poisoning on human health had spread outside of the central area and could have still been caused until recently. By using the frequency of fish ingestion as an indirect indication of methylmercury exposure, a dose-response relationship was confirmed for methylmercury pollution in Minamata. The latency period from exposure of mercury to the onset of symptoms was much longer than previously thought, and the latency period increased as the exposure levels decreased. Further investigations on the health effects of methylmercury over a wider area, covering longer periods, and on a wider range of signs and symptoms must continue in the future.

Author Contributions: S.T. and T.F. conceived and designed this survey; S.T., T.F., Y.K. and S.-i.S. performed the survey; S.T. analyzed the data; T.Y. contributed analysis tools; S.T. wrote the paper.

Funding: This survey received no external funding for carrying out this study.

Acknowledgments: We: the authors, would like to express our thanks to the late Masazumi Harada of the Open Research Center for Minamata Studies, Kumamoto Gakuen University, for his involvement in helping us carry out this survey. We would also like to thank all medical doctors, nurses, health care professionals, and other participants who cooperated in the execution of this survey. All the participants volunteered to take part in this survey.

Conflicts of Interest: The authors declare no conflict of interest.

References

1. Harada, M. Minamata disease: Methylmercury poisoning in japan caused by environmental pollution. *Crit. Rev. Toxicol.* **1995**, *25*, 1–24. [CrossRef] [PubMed]
2. Yorifuji, T.; Tsuda, T.; Inoue, S.; Takao, S.; Harada, M.; Kawachi, I. Critical appraisal of the 1977 diagnostic criteria for minamata disease. *Arch. Environ. Occup. Health* **2013**, *68*, 22–29. [CrossRef] [PubMed]
3. Uchino, M. *Transition of Clinical Picture of Minamata Disease: Analysis of Neurological Symptom of Physical Examination for Certification of Minamata Disease*; Japan Public Health Association: Tokyo, Japan, 1997; pp. 72–76. (In Japanese)
4. Minamata City. Compensastion & relief of victims—Health measures for residents. In *Minamata Disease—Its History and Lessons*; Minamata City Planning Division: Minamata, Japan, 2000.
5. Takizawa, Y.; Osame, M. A brief introduction to Minamata disease. In *Methylmercury Poisoning in Minamata and Niigata, Japan*; Takizawa, Y., Osame, M., Eds.; Japan Public Health Association: Tokyo, Japan, 2001; pp. 3–4.
6. Takaoka, S.; Kawakami, Y.; Fujino, T.; Oh-ishi, F.; Motokura, F.; Kumagai, Y.; Miyaoka, T. Somatosensory disturbance by methylmercury exposure. *Environ. Res.* **2008**, *107*, 6–19. [CrossRef] [PubMed]
7. Shigenaga, K.; Sato, K.; Takeuchi, T. Pathological changes of pancreatic islets in Minamata disease. *Kumamoto Igakkai Zasshi* **1974**, *48*, 189–198. (In Japanese)
8. Harada, M. Clinical and epidemiological studies of Minamata disease in 16 years after onset. *Shinkeikenkyuu no Shinpo* **1972**, *16*, 870–880. (In Japanese)

9. Tatetsu, S.; Kiyota, K.; Harada, M.; Miyakawa, T. Study on health effect on shiranui seacoast residents by ingestion of fish and shellfish polluted by organic mercury. In *Epidemiological, Clinical, and Pathological Study on Minamata Disease 10 Years after the Outbreak*; Medical School of Kumamoto University: Kumamoto, Japan, 1993; pp. 48–87. (In Japanese)

10. Fujino, T. Clinical and epidemiological studies on chronic Minamata disease part I: Study on Katsurajima island. *Kumamoto Med. J.* **1994**, *44*, 139–155.

11. Kinjo, Y.; Higashi, H.; Nakano, A.; Sakamoto, M.; Sakai, R. Profile of subjective complaints and activities of daily living among current patients with minamata disease after 3 decades. *Environ. Res.* **1993**, *63*, 241–251. [CrossRef] [PubMed]

12. Sugiura, A. Health conditions among fishernen living in the Minamata disease prevalent area. *Jpn. J. Public Health* **1994**, *41*, 428–440. (In Japanese)

13. Ninomiya, T.; Ohmori, H.; Hashimoto, K.; Tsuruta, K.; Ekino, S. Expansion of methylmercury poisoning outside of Minamata: An epidemiological study on chronic methylmercury poisoning outside of Minamata. *Environ. Res.* **1995**, *70*, 47–50. [CrossRef] [PubMed]

14. Takeuchi, T. Neuropathology of Minamata disease: Especially at the chronic stage. In *Neurotoxicology*; Roizin, L., Shiraki, H., Grcevie, N., Eds.; Raven Press: New York, NY, USA, 1977; pp. 235–246.

15. Ceccatelli, S.; Dare, E.; Moors, M. Methylmercury-induced neurotoxicity and apoptosis. *Chem. Biol. Interact.* **2010**, *188*, 301–308. [CrossRef] [PubMed]

16. Uchino, M.; Araki, S. Clinical features of chronic Minamata disease (organic mercury poisoning). Analysis of the neurological findings in the most recent 100 cases. *Rinsho Shinkeigaku* **1984**, *24*, 235–239. (In Japanese) [PubMed]

17. Central Council for Environmental Pollution Control. *About Future Measures to Minamata Disease*; The Environmental Agency: Bristol, UK, 1991. (In Japanese)

18. Igata, A. Epidemiological and clinical features of Minamata disease. In *Advances in Mercury Toxicology*; Suzuki, T., Imura, N., Clarkson, T., Eds.; Plenum Press: New York, NY, USA, 1991; pp. 439–457.

19. Bakir, F.; Damluji, S.F.; Amin-Zaki, L.; Murtadha, M.; Khalidi, A.; Al-Rawi, N.Y.; Tikriti, S.; Dahahir, H.I.; Clarkson, T.W.; Smith, J.C.; et al. Methylmercury poisoning in iraq. *Science* **1973**, *181*, 230–241. [CrossRef] [PubMed]

20. Rice, D.C. Evidence for delayed neurotoxicity produced by methylmercury. *Neurotoxicology* **1996**, *17*, 583–596. [PubMed]

21. Evans, H.L.; Garman, R.H.; Weiss, B. Methylmercury: Exposure duration and regional distribution as determinants of neurotoxicity in nonhuman primates. *Toxicol. Appl. Pharmacol.* **1977**, *41*, 15–33. [CrossRef]

22. Weiss, B.; Clarkson, T.W.; Simon, W. Silent latency periods in methylmercury poisoning and in neurodegenerative disease. *Environ. Health Perspect.* **2002**, *110* (Suppl. 5), 851–854. [CrossRef] [PubMed]

23. Julvez, J.; Yorifuji, T.; Choi, A.L.; Grandjean, P. Chapter 2 epidemiological evidence on methylmercury neurotoxicity. In *Methylmercury and Neurotoxicity*; Ceccatelli, S., Aschner, M., Eds.; Springer: New York, NY, USA, 2012; pp. 14–35.

24. Karagas, M.R.; Choi, A.L.; Oken, E.; Horvat, M.; Schoeny, R.; Kamai, E.; Cowell, W.; Grandjean, P.; Korrick, S. Evidence on the human health effects of low-level methylmercury exposure. *Environ. Health Perspect.* **2012**, *120*, 799–806. [CrossRef] [PubMed]

Correction

toxics

MDPI

Correction: Takaoka, S., et al. Survey of the Extent of the Persisting Effects of Methylmercury Pollution on the Inhabitants around the Shiranui Sea, Japan. *Toxics* 2018, 6, 39

Shigeru Takaoka [1,*], Tadashi Fujino [2], Yoshinobu Kawakami [3], Shin-ichi Shigeoka [3] and Takashi Yorifuji [4]

1   Kyoritsu Neurology and Rehabilitation Clinic, 2-2-28 Sakurai-cho, Minamata 867-0045, Japan
2   Kikuyou Hospital, 5587 Haramizu, Kikuyou 869-1102, Japan; tds-fujino@jcom.zaq.ne.jp
3   Minamata Kyoritsu Hospital, 2-2-12 Sakurai-cho, Minamata 867-0045, Japan; mkkawa@fsinet.or.jp (Y.K.); shigeoka@mk-kyouritu.com (S.-i.S.)
4   Department of Human Ecology, Graduate School of Environmental and Life Science, Okayama University, 3-1-1 Tsushima-naka, Kita-ku, Okayama 700-8530, Japan; yorichan@md.okayama-u.ac.jp
*   Correspondence: stakaoka@x.email.ne.jp; Tel.: +81-966-63-6835

Received: 19 March 2019; Accepted: 22 March 2019; Published: 11 April 2019

The authors wish to make the following corrections to this paper [1]:

1. Page 1, Abstract, lines 12–15: Period of time changed to match time between 1968 and the year in which our data was taken:

"Our data indicates that Minamata disease had spread outside of the central area and could still be observed recently, almost 50 years after the Chisso Company's factory had halted the dumping of mercury polluted waste water back in 1968." should be changed to "Our data indicates that Minamata disease had spread outside of the central area, and could still be observed recently, almost 40 years after the Chisso Company's factory had halted the dumping of mercury polluted waste water back in 1968.".

2. Page 3, paragraph 2, lines 5–7: Restructuring of sentence to make it clearer:

"Applicants seeking recognition were being socially discriminated against and the lack of a comprehensive pollution survey meant that many residents with health problems had not sought diagnosis." should be changed to "Due to social discrimination and the lack of a comprehensive pollution survey, residents with health problems were reluctant to seek a diagnosis.".

3. Page 4, paragraph 2, lines 2–4: Incorrect word replaced by correct one:

"We performed this survey in order to research the prevalence of signs and symptoms as well as the geometrical and chronological spread of health problems caused by methylmercury." should be changed to "We performed this survey in order to research the prevalence of signs and symptoms, as well as the geographical and chronological spread of health problems caused by methylmercury.".

4. Page 6, paragraph 1, lines 5–6: Missing word inserted:

"Those who had been born or had moved to the polluted on or after 1 January 1969 were classified under the third category (BA1968: $n = 30$, M/F = 21/9, Age = 37.4 ± 2.3)." should be changed to "Those who had been born or had moved to the polluted area on or after 1 January 1969 were classified under the third category (BA1968: $n = 30$, M/F = 21/9, Age = 37.4 ± 2.3).", adding the word 'area'.

5. Page 6, paragraph 2, lines 2–5: Rearrange order at the end of the paragraph:

"To evaluate this group, we selected 88 out of 227 subjects whose age was lower than 49 from the Control Area (M/F = 40/48, Age = 37.5 ± 6.0), and 84 out of 786 exposed subjects in the designated area who were born after 31 December 1968 (M/F = 44/40, Age = 44.8 ± 2.3) and whose age was lower than 49 from the four exposed groups." should be changed to "To evaluate this group, we selected 88 out of 227 subjects, whose age was lower than 49, from the Control Area (M/F = 40/48, Age = 37.5 ± 6.0) and 84 out of 786 exposed subjects in the designated area, who were born after 31 December 1968 and whose age was lower than 49, from the four exposed groups (M/F = 44/40, Age = 44.8 ± 2.3).".

6. Page 8, 2.6.2, lines 2–3: Added missing words to clarify better:

"To evaluate the severity of the neurological signs, we added (a) mark(s) to positive signs and symptoms, and we calculated the total score in the exposed four groups." should be changed to "To evaluate the severity of the neurological signs and symptoms, we added (a) mark(s) to positive signs and symptoms, and we calculated the total score in the exposed four groups and the control group.", adding 'and symptoms', as well as 'and the control group'.

7. Page 21, Section 3.5.2, line 1: Remove "other":

"The frequency of fish ingestion was closely related to the onset year of other symptoms (Table 13, Figure 9)." should be changed to "The frequency of fish ingestion was closely related to the onset year of the symptoms (Table 13, Figure 9).".

8. Page 23, Table 14: Table title shortened:

"Table 14. Score of signs and onset of symptoms in each area." should be changed to "Table 14. Score of signs and onset of symptoms.".

9. Page 27, Figure 17: Changed "symptoms" to "signs":

"Figure 17. Prevalence of neurological signs in subjects with and without sensory disturbance. Except for the prevalence of sensory disturbance, that of other symptoms was lower in subjects without sensory disturbance than in subjects with sensory disturbance but was generally higher than the Control Area. The prevalence patterns were similar in exposed subjects with and without sensory disturbance except for the prevalence of sensory disturbance." should be changed to "Figure 17. Prevalence of neurological signs in subjects with and without sensory disturbance. Except for the prevalence of sensory disturbance, that of other signs was lower in subjects without sensory disturbance, than in subjects with sensory disturbance, but was generally higher than the Control Area. The prevalence patterns were similar in exposed subjects with and without sensory disturbance except for the prevalence of sensory disturbance.".

10. Page 32, Figure 18, Missing word inserted:

"Figure 18. Prevalence of symptoms when comparing nine questions common to all five studies. The prevalence of all characteristic symptoms for Minamata disease was very. In this study, we used the prevalence of "sensory numbness in both hands" instead of "numbness of hands and feet."" should be changed to "Figure 18. Prevalence of symptoms when comparing nine questions common to all five studies. The prevalence of all characteristic symptoms for Minamata disease was very high. In this study, we used the prevalence of "sensory numbness in both hands" instead of "numbness of hands and feet."", thus inserting 'high' as in 'very high' in the first sentence.

11. Page 32, paragraph 2, lines 1–2: The age range data used for ADL comparison was 60–69. This resulted in a change for the value for bodily hygiene from 10.6% to 7.0%. The scientific results, however, remain the same:

"Although activities of daily living (ADL) of these patients decreased from 60 years old, 9.1% still needed assistance in eating, 11.6% in bodily hygiene and 10.6% in using the toilet [11]." should be changed to "Although activities of daily living (ADL) of these patients decreased from 60 years old, 9.1% still needed assistance in eating, 11.6% in bodily hygiene, and 7.0% in using the toilet in the age range of 60–69 [11].", where the percentage of those using the toilet has been changed from 10.6% to 7.0%, and the words 'in the age range of 60–69' have also been added at the end.

12. Page 32, paragraph 3, lines 5–6: Referenced table numbers corrected:

"Tables 2 and 4, Figures 2 and 3 show that prevalence of specific symptoms as well as that of non-specific symptoms became higher through methylmercury exposure." should be changed to "Tables 3 and 5, Figures 2 and 3 show that prevalence of specific symptoms as well as that of non-specific symptoms, became higher through methylmercury exposure.".

13. Page 33, 4.3, lines 7–8: The following sentence should be removed, as its meaning is incorrect:

This is the first study that dose-response effects were observed in methylmercury poisoning in Japan.

14. Page 34, 4.6, paragraph 3, lines 1–2: Reworded to clear up possible ambiguity:

"It is difficult to determine whether subjects of BA1968 (as displayed in Figure 14) had developed their symptoms due to continued exposure after 1968 or if they were late developing symptoms." should be changed to "We can understand that subjects of BA1968 (as displayed in Figure 14) had developed their symptoms due to continued exposure after 1968. But it is impossible to determine whether subjects who were born before BA1968 and had developed their symptoms after 1968 had developed their symptoms due to continued exposure after 1968 or if they were late developing symptoms resulting from exposure before 1968.".

15. Page 34, Section 4.7, lines 3–5 Added reference to some international studies:

"Outside of Japan, epidemiological studies have not seen such extreme neurological signs, but other more mild or latent neurocognitive and behavioral symptoms [23,24]." should be changed to "Outside of Japan, epidemiological studies with such extreme neurological signs are rare except for Iraq, Canada, and so on, but many other more mild or latent neurocognitive and behavioral symptoms have been reported [23,24].".

16. Page 35, 4.8, paragraph 2, line 1: Reworded to clear up possible ambiguity:

"Secondly, the lifestyle and occupations were not the same among the four groups." should be changed to "Secondly, the lifestyle and occupations were not the same between exposed and control groups.".

17. Page 36, reference 12. Misspelling corrected:

"12. Sugiura, A. Health conditions among fisheren living in the Minamata disease prevalent area. *Jpn. J. Public Health* **1994**, *41*, 428–440. (In Japanese)" should be changed to "12. Sugiura, A. Health conditions among fishermen living in the Minamata disease prevalent area. *Jpn. J. Public Health* **1994**, *41*, 428–440. (In Japanese)".

18. Page 36, reference 17. Incorrect reference to "Bristol, UK" removed:

"17. Central Council for Environmental Pollution Control. *About Future Measures to Minamata Disease*; The Environmental Agency: Bristol, UK, 1991. (In Japanese)" should be changed to "17. Central Council for Environmental Pollution Control. *About Future Measures to Minamata Disease*; The Environmental Agency: Tokyo, Japan, 1991. (In Japanese)".

None of the above changes affect the scientific results. The manuscript will be updated and the original will remain online on the article webpage. We apologize for any inconvenience caused to our readers.

Reference

1. Takaoka, S.; Fujino, T.; Kawakami, Y.; Shigeoka, S.-I.; Yorifuji, T. Survey of the Extent of the Persisting Effects of Methylmercury Pollution on the Inhabitants around the Shiranui Sea, Japan. *Toxics* **2018**, *6*, 39. [CrossRef] [PubMed]

Article

Health Impacts and Biomarkers of Prenatal Exposure to Methylmercury: Lessons from Minamata, Japan

Mineshi Sakamoto [1,2,*], Nozomi Tatsuta [2], Kimiko Izumo [1], Phuong Thanh Phan [3], Loi Duc Vu [4], Megumi Yamamoto [1], Masaaki Nakamura [1], Kunihiko Nakai [2] and Katsuyuki Murata [5]

[1] National Institute for Minamata Disease, Minamata, Kumamoto 867-0008, Japan; izumo@nimd.go.jp (K.I.); yamamoto@nimd.go.jp (M.Y.); nakamura@nimd.go.jp (M.N.)
[2] Development and Environmental Medicine, Tohoku University Graduate School of Medicine, Miyagi 980-8575, Japan; nozomi@med.tohoku.ac.jp (N.T.); winestem@med.akita-u.ac.jp (K.N.)
[3] Faculty of Chemistry, Thai Nguyen University of Sciences, Thai Nguyen 250000, Vietnam; phuongqtdhkh@gmail.com
[4] Institute of Chemistry, Vietnam Academy of Science and Technology, Hanoi 100000, Vietnam; ducloi@ich.vast.vn
[5] Department of Environmental Health Sciences, Akita University School of Medicine, Akita 010-8543, Japan; satoc@med.tohoku.ac.jp
* Correspondence: sakamoto@nimd.go.jp; Tel.: +81-966-63-3111 (ext. 312); Fax: +81-966-61-1145

Received: 13 June 2018; Accepted: 31 July 2018; Published: 3 August 2018

Abstract: The main chemical forms of mercury are elemental mercury, inorganic divalent mercury, and methylmercury, which are metabolized in different ways and have differing toxic effects in humans. Among the various chemical forms of mercury, methylmercury is known to be particularly neurotoxic, and was identified as the cause of Minamata disease. It bioaccumulates in fish and shellfish via aquatic food webs, and fish and sea mammals at high trophic levels exhibit high mercury concentrations. Most human methylmercury exposure occurs through seafood consumption. Methylmercury easily penetrates the blood-brain barrier and so can affect the nervous system. Fetuses are known to be at particularly high risk of methylmercury exposure. In this review, we summarize the health effects and exposure assessment of methylmercury as follows: (1) methylmercury toxicity, (2) history and background of Minamata disease, (3) methylmercury pollution in the Minamata area according to analyses of preserved umbilical cords, (4) changes in the sex ratio in Minamata area, (5) neuropathology in fetuses, (6) kinetics of methylmercury in fetuses, (7) exposure assessment in fetuses.

Keywords: methylmercury; kinetics; toxicity; fetus; exposure assessment

---

1. Methylmercury Toxicity

Mercury has been used by humans for centuries due to its unique physical and chemical properties. However, the different chemical forms of mercury, which include elemental mercury (Hg^0), inorganic divalent mercury (Hg^{2+}), and organic mercury (mainly methylmercury, CH_3Hg^+), cause a variety of toxic effects [1,2]. Differences in exposure sources, the affected organs, toxic effects, and metabolism are seen among the various chemical forms of mercury. For example, the general population is exposed to small amounts of elemental mercury due to its use in dental amalgams. On the other hand, workers at artisanal small-scale gold mining sites and gold shops in Amazon River regions can be exposed to high levels of elemental mercury, as they often have to heat gold-mercury amalgams to evaporate the mercury and obtain the gold [2]. In humans, minimal amounts of liquid elemental mercury are absorbed from the digestive tract, and so liquid elemental mercury does not cause acute toxicity, even when the liquid mercury in a thermometer is accidentally ingested. However, problems can arise when liquid mercury is heated and bursts into the surrounding air [2]. A large proportion of inhaled gaseous elemental mercury (approximately 80%) is absorbed into the blood via the lungs, and, as an uncharged and therefore lipid-soluble substance, it can easily pass through the blood-brain barrier. With time, the gaseous elemental mercury in the patient's brain is oxidized to inorganic divalent mercury and causes damage to the brain, and the inorganic mercury also accumulates in the kidneys, where it causes renal damage [2]. The amount of inorganic mercury absorbed through the digestive tract is comparatively low (5–10%) [3]. However, the intake of a large amount of inorganic mercury compounds, such as mercury (II) chloride, in cases of accidental ingestion or ingestion with suicidal intent causes digestive tract and kidney disorders, which can result in death [3]. In the environment, a part of the divalent mercury can be changed to methylmercury by some micro-organisms and sunlight. It then bioaccumulates in fish and marine mammals which exhibit elevated, food-web dependent methylmercury levels [4]. Most human methylmercury exposure occurs from the consumption of fish and seafood. Methylmercury is readily absorbed by the digestive tract (>90% is absorbed) [5]. In addition, it exhibits high affinity for sulfhydryl groups [1], and some methylmercury combines with L-type cysteine to form L-cysteine-methylmercury conjugates, which have similar chemical structures to methionine, an essential amino acid [6]. The conjugates are then distributed to all tissues, including the brain (via the blood brain barrier), as they are treated like L-type neutral amino acids [7,8]. In the epidemics of marked methylmercury intoxications in Minamata [9,10], Japan and Iraq [11], the brain was the organ that was most severely affected, particularly in that the developing brains of fetuses were damaged [12,13].

2. History and Background of Minamata Disease

The epidemic of methylmercury intoxication that occurred in Minamata area, Kumamoto Prefecture, Japan, is known as "Minamata disease". It was the first experience of severe methylmercury poisoning caused by anthropogenic environmental pollution [10]. The causative agent, methylmercury, was produced from inorganic mercury as a byproduct of the process used to manufacture acetaldehyde at Chisso Co., Ltd. (Kumamoto, Japan), in Minamata City, and it was directly discharged into Minamata Bay [10]. People who consumed a large amount of fish and shellfish that had been contaminated with methylmercury from Minamata Bay developed symptoms of methylmercury toxicity. The first patient to suffer from neurological symptoms was reported in May 1956. Kumamoto University soon started to investigate the cause of this outbreak, and in March 1957 they reported that they suspected that this disease was a type of heavy metal poisoning transmitted via fish and shellfish consumption [10]. However, because this was the first case of methylmercury poisoning due to environmental pollution it took many years for the cause and effect relationship to be fully elucidated. In the meantime, the number of patients with Minamata disease started to rapidly increase, mainly in the fishing village of Minamata [10]. Therefore, people were afraid that the disease was "a strange contagious disease" with an unknown cause. In 1958, Chisso Co., Ltd., moved their effluent outlet from Minamata Bay to the Minamata River in an effort to ameliorate the epidemic taking place in the areas near Minamata. However, this resulted in the disease spreading to the areas surrounding Minamata City. In addition, a second epidemic of methylmercury poisoning, so-called "Niigata Minamata disease", occurred in 1965. This outbreak was caused by methylmercury derived from the same acetaldehyde production process as was responsible for the epidemic in Minamata City. Chisso Co., Ltd. stopped acetaldehyde production in May 1969. In September 1969, almost 12 years after the first case of Minamata disease was encountered, the Japanese government officially announced that the causative agent of Minamata disease was methylmercury, which had been discharged from the above-mentioned chemical plants.

3. Methylmercury Pollution in the Minamata Area According to Analyses of Preserved Umbilical Cords

Unfortunately, no human or biota samples were collected from the Minamata area during the period of severe methylmercury pollution, and so methylmercury exposure levels could not be determined. Therefore, the time-course and regional distribution of methylmercury pollution in the Minamata area were unknown. However, Japanese people have an ancient custom of preserving their children's umbilical cords as mementos, and so it was possible to assess methylmercury exposure levels in the Minamata area by analyzing the preserved umbilical cords of children born in the region, which in turn provided a reliable estimate of the time-course of the changes in methylmercury pollution [14,15]. A study by Nishigaki and Harada (1975) achieved a breakthrough in that it revealed that the Minamata inhabitants' methylmercury exposure levels peaked in the first five years of the 1950s, and then the methylmercury levels of the preserved umbilical cord tissue samples decreased according to the decline of acetaldehyde production in the Minamata area. Sakamoto et al. [16] analyzed a total of 325 umbilical cord samples, including 124 newly collected samples and the 164 samples collected during the studies published by Harada et al. [15,17,18]. Figure 1 shows the methylmercury concentrations of individual preserved umbilical cords (μg/g dry weight) from the Minamata area and the annual level of acetaldehyde production at the time the samples were collected. Elevated methylmercury concentrations ($\geq$1 μg/g) were mainly observed in the inhabitants born from 1947 to 1968. The peak methylmercury concentrations ($\geq$2 μg/g) were mainly observed during the period from 1955 to 1959, when the typical fetal-type Minamata disease patients were born [13], and a reduction in the frequency of the male sex was detected in the Minamata area by Sakamoto et al. [16]. The residents' methylmercury concentrations started to decrease along with the decline in acetaldehyde production, which ceased in 1968. After 1968, no individuals with elevated methylmercury concentrations ($\geq$1 μg/g) were encountered. These unique retrospective studies of the methylmercury concentrations of preserved umbilical cord samples revealed not only the historical time-course of methylmercury pollution, but also its regional distribution, in the Minamata area. Sakamoto et al. [19] also calculated a conversion factor for converting methylmercury concentrations in dry weight of cord tissue to the equivalent maternal hair (0–1 cm from the scalp) level (conversion factor: 24.09). Preserved umbilical cord tissue would also be useful for retrospective dose-response studies of methylmercury exposure and the occurrence of fetal-type Minamata disease in Japan.

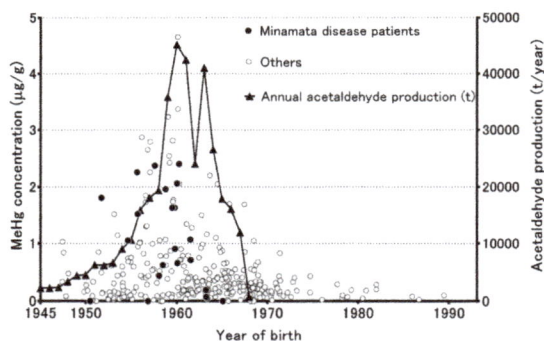

Figure 1. Individual methylmercury concentrations of preserved umbilical cords (μg/g dry weight) from the Minamata area and the amount of annual acetaldehyde production.

4. Changes in the Sex Ratio in Minamata City

Several decades ago, skewed sex ratios at birth due to hazardous chemicals became a matter of international concern, especially the skewed sex ratio caused by dioxin contamination in Seveso, Italy [20]. As mentioned above, prenatal methylmercury exposure has much stronger effects on fetuses than on their mothers. Therefore, Sakamoto et al. examined the sex ratio at birth in Minamata area to evaluate the effects of severe methylmercury exposure [16]. In four of the five years from 1955 to 1959, unexpectedly low numbers of males were born in Minamata City. Furthermore, a dose-dependent (based on the estimated environmental level of methylmercury) reduction in the frequency of the male sex was observed in that period (male sex:total population ratio; 0.515 for Kumamoto City (the control), >0.492 for Minamata City, >0.459 for the area in which Minamata disease was most prevalent, and >0.382 among the families of fishermen). The lowest male:total population ratio (0.393) was observed among the offspring of mothers with Minamata disease [16]. However, no reduction in the frequency of the male sex was observed in the cases in which only the father was affected by Minamata disease. This phenomenon was suspected to have been caused by maternal methylmercury exposure leading to methylmercury having direct effects on fetuses. In addition, the proportion of male stillbirths in the city increased to 173 males/100 females (male proportion: 0.634) when the methylmercury pollution was at its most severe, indicating that more males were lost at the fetal stage. The increased proportion of stillborn males in Minamata City partly explains the lower proportion of males in the abovementioned study. However, little is known about the sex-related differences in the susceptibility of fetuses against the lethal effects of methylmercury.

5. Neuropathology in Fetuses

Among the adults with Minamata disease that were exposed to high levels of methylmercury, neuronal degeneration was observed from medical autopsies, predominantly in certain areas of the cerebral cortex (the parietal, occipital, and temporal lobes), cerebellum, and peripheral nerves [9]. Figure 2a shows the lesion distribution among adult cases of Minamata disease. The main symptoms exhibited by these patients included sensory disturbances in the distal parts of the extremities followed by ataxia, concentric contraction of the visual field, impairment of gait and/or speech, muscle weakness, tremors, abnormal eye movement, and hearing impairment [21], which mainly reflected the areas of the brain that suffered nerve damage [9]. In Iraq, a large-scale epidemic of methylmercury poisoning occurred in 1972–1973 after wheat seeds were disinfected with methylmercury [11]. This outbreak affected more than 6000 people and resulted in 400 deaths. The main symptoms were similar to those of Minamata disease. A study conducted in Iraq [11] showed that in adult cases of methylmercury poisoning the estimated mercury body burden thresholds (mg) at diagnosis for various symptoms were as follows: abnormal sensory perception, about 25 mg (equivalent to a mercury blood concentration of 250 µg/L); ataxia, about 50 mg; articulation disorders, about 90 mg; hearing loss, about 180 mg; death, >200 mg.

In Minamata City, a high incidence of cerebral palsy was observed from 1955 to 1959, when the most severe cases of Minamata disease occurred. The incidence rate was 5.8%, which was much higher than the normal incidence rate (0.2–0.64%) in Japan [10]. The study group of Kumamoto University concluded that they were fetal-type Minamata disease patients, who were exposed to methylmercury via placenta [10]. The symptoms observed in fetal-type Minamata disease patients (22 typical severe cases) were mental retardation, inability of walking by oneself, disturbances of coordination, speech, chewing and swallowing, and increased muscle tone, which are similar to the symptoms of cerebral palsy [10]. Histopathological examinations of Japanese fetal-type Minamata disease patients revealed widespread and severe neuronal degeneration in the central nervous system [18]. Figure 1b shows the lesion distribution fetal-type cases of Minamata disease. On the other hand, their mothers had mild or no manifestations of methylmercury poisoning [13]. In Iraq, the children that were most severely affected by methylmercury poisoning manifested with severe sensory impairments, general paralysis, hyperactive reflexes, and/or impaired mental development [11]. The world Health Organization suggested that Iraqi data implied that a peak maternal hair mercury level of 10–20 µg/g is associated with a 5% risk of neurological disorders [5].

In an animal experiment using rats, distinct patterns of neuronal degeneration were observed [22] by methylmercury administration at various stages of brain development. For example, neonatal rats administered methylmercury for 10 days from postnatal day (PD) 2 showed minimal damage in the hippocampus and brainstem nuclei, young rats administered 10 days from PD 15 showed neuronal degeneration in the cerebral cortex, striatum, and red nucleus, while adult rats from PD 60 showed severe lesions in the cerebellum and dorsal root ganglia. Moreover, neonatal rats exposed to methylmercury for prolonged period (>30 days) from PD 2 showed the widespread neuronal degeneration in the cerebral neocortex and spinal sensory ganglia [23]. These rat experiments indicated that the distribution pattern of methylmercury-induced brain lesions changes depending on the stage of brain development at which methylmercury exposure occurs, and the widespread neuronal degeneration in human fetal-type Minamata disease could be caused by prolonged methylmercury exposure during brain growth.

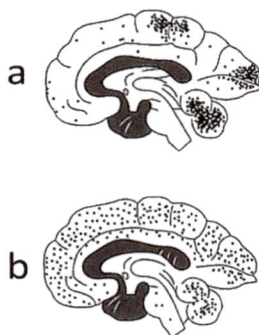

Figure 2. Comparisons of the distributions of lesions among adult and fetal-type cases of Minamata disease. Note: The distributions of degenerated neurons in adult (**a**) and fetal patients (**b**) are shown. (Modified from Reference [9] with permission).

6. Kinetics of Methylmercury in Fetuses

Fetuses depend on their mothers for nutrients, including amino acids, fatty acids, and vitamins. However, they can also be exposed to methylmercury through the maternal consumption of fish and shellfish. Furthermore, some animal studies have shown suggested that methylmercury is actively transferred from mother to fetus via placental amino acid transport systems [24–26]. In humans, it was reported that cord blood contains higher concentrations of methylmercury than maternal blood [27–31]. Especially, Stern and Smith [31] summarized the data of the cord blood/maternal blood methylmercury ratio from 10 reports published from 1975 to 2000. Table 1 shows the total maternal and cord blood mercury concentrations recorded in various study populations from the recent papers which were published after 2016. Although the cord blood/maternal ratio varied from 1.03 to 2.04, all the data indicated that cord blood mercury levels were higher than those of maternal blood as summarized by Stern and Smith [31].

Table 1. Total maternal and cord blood mercury concentrations in various study populations.

Study Site	Measure		Maternal Blood Hg	Cord Blood Hg	Cord Blood/Maternal Blood Ratio	Sampling Years and (Published Year) References
Ten cites, Canada	T-Hg	µg/L	0.562 (*n* = 1673)	0.802 (*n* = 1419)	1.43	2008–2011 (2016) [32]
Laizhou By, China	T-Hg	µg/L	0.72 (*n* = 410)	1.20 (*n* = 410)	1.67	2010–2012 (2016) [33]
Busan, Korea	T-Hg	µg/L	3.12 (*n* = 127)	5.46 (*n* = 127)	1.75	2009–2010 (2016) [34]
Tong Gang, Taiwan	T-Hg	µg/L	2.24 (*n* = 145)	2.30 (*n* = 145)	1.03	2010–2011 (2017) [35]
Tokyo, Japan	T-Hg	µg/L	4.97 (*n* = 334)	10.15 (*n* = 334)	2.04	2010–2012 (2018) [36]
Kumamoto, Japan	T-Hg	ng/g	3.79 (*n* = 54)	7.26 (*n* = 54)	1.92	2006–2007 (2018) [37]

In developing fetuses, the brain is sensitive to methylmercury exposure [5]. Both the high sensitivity of the developing brain to methylmercury [38] and the increased methylmercury accumulation in the blood and brain [26] of fetuses are recognized toxicological features of methylmercury. Consequently, the effects of dietary seafood intake in pregnant women remain an important public health issue, especially in populations that consume large quantities of fish and sea mammals, such as toothed whales and seals. Mercury levels in pregnant women can be affected by fish consumption patterns such as the amount, species, frequency, and seasons of the fish consumption during pregnancy. Consequently, the comparison among studies is not always easy [39].

7. Exposure Assessment in Fetuses

Since the brain is the organ that is most at risk from methylmercury, the biomarkers used to determine human methylmercury exposure levels should reflect the methylmercury concentration in the brain. In humans, methylmercury has an average biological half-life of approximately 70 days (whole body) [5]. Generally, the amount retained in the body reaches an equilibrium during constant methylmercury intake, e.g., from seafood consumption. Animal experiments have indicated that the ratio of the blood mercury concentration to the brain mercury concentration remains constant under steady state conditions. Therefore, the mercury concentration in the blood/red blood cells is a good biomarker for assessing methylmercury exposure [5]. The mercury concentration in hair reflects the blood methylmercury concentration during hair formation and is frequently used as a biomarker for evaluating methylmercury exposure [5]. Although analyses of hair mercury concentrations are affected by a number of variables, such as the hair's growth rate, density, color, waving, external contamination, and any permanent treatments [5], segmental analysis of maternal hair is able to provide time-course information because the average hair growth rate is commonly assumed to be about 1 cm per month [40,41]. On the other hand, cord blood circulates in the fetal body and can directly reflect the methylmercury concentrations in fetal organs, including the fetal brain, at birth [41]. In addition, a number of studies have employed toenail and/or fingernail mercury concentrations as biomarkers for assessing methylmercury exposure [42–45]. In most of these studies, toenails rather than fingernails were preferred, because toenails are often less contaminated than fingernails, especially among dental personnel and gold miners, who handle mercury amalgams [2].

The organ that is most affected by methylmercury exposure during gestation is the fetal brain. For this reason, biomarkers that reflect fetal methylmercury exposure during gestation are very important for predicting the effects of methylmercury on child development. In a study conducted in the Faroe Islands, the cord blood mercury concentration was the preferred biomarker for evaluating methylmercury exposure, whereas in a study carried out in the Seychelles, the maternal hair mercury concentration was used as the only biomarker of fetal exposure. Umbilical cord tissue has also been used to determine fetal methylmercury exposure levels in some studies [14,46]. In addition, maternal mercury levels in fingernails and toenails at parturition showed strong correlations with those in cord blood [45]. Figure 3 shows the correlation coefficients (r) for the relationships among various biomarkers of methylmercury exposure at parturition, which were obtained from our previous studies [30,45]. All of the biomarkers, including maternal blood, maternal hair, cord blood, maternal nails, the placenta, and cord tissue showed strong correlations with each other (r: >0.70). This suggests that all of the examined biomarkers are useful for assessing the prenatal exposure of fetuses to methylmercury.

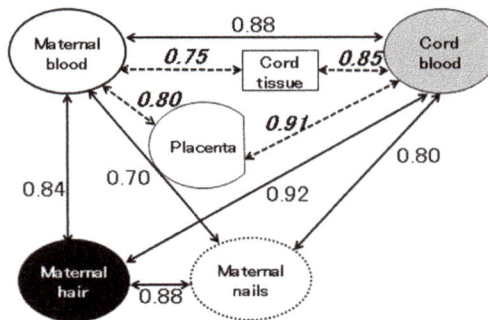

Figure 3. Correlation coefficients (r) for the relationships among biomarkers of methylmercury exposure at parturition. Note: Red blood cells were used to calculate the correlations between the levels of methylmercury in blood, the placenta, and cord tissue (- - -) [30]. Whale blood was used to calculate the correlations between the methylmercury levels of blood, hair, and nails (—) [45].

8. Summary

The United Nations Environment Programme (UNEP) agreed to develop a global legally-binding instrument on Hg in 2013 as the Minamata Convention on Mercury [4], and the treaty entered into force in August of 2017. UNEP concluded that the rapid increase in historical environmental Hg levels began during the industrial revolution in the 19th century. The impact of increased Hg will appear as increased methylmercury levels in the marine environment, especially in fish and marine mammals, and finally in humans who consume marine products. Among the human populations, fetuses who bear the next generation are at the highest risk to the methylmercury exposure. Therefore, we need an effort to reduce the anthropogenic emissions/releases of mercury into the environment, "recognizing to reflect the lessons learned from Minamata disease".

Author Contributions: This publication was carried out by all the authors; M.S. conceived the idea and wrote the draft; N.T., K.I., P.T.P., L.D.V., M.Y., M.N., K.N., and K.M. managed literature search and scrutinized the draft.

Funding: This research was funded in part by a grant from the Japan Ministry of the Environment. The findings and conclusions of this article are solely the responsibility of the authors and not represent the official views of the above agency.

Conflicts of Interest: The authors declare no conflict of interest.

References

1. NRC (National Research Council). *Toxicological Effects of Methylmercury*; Academic Press: Washington, DC, USA, 2000.
2. UNIDO (United Nations Industrial Development Organization). *Protocols for Environmental and Health Assessment of Mercury Released by Artisanal and Small-Scale Gold Miners*; United Nations Industrial Development Organization: Vienna, Austria, 2008.
3. WHO (World Health Organization). *Inorganic Mercury. Environmental Health Criteria 118*; World Health Organization: Geneva, Switzerland, 1991.
4. UNEP (United Nations Environment Programme). *Global Mercury Assessment 2013: Sources, Emissions, Releases and Environmental Transport*; UNEP Chemicals Branch: Geneva, Switzerland, 2013.
5. WHO (World Health Organization). *Methylmercury. Environmental Health Criteria 101*; World Health Organization: Geneva, Switzerland, 1990.
6. Yin, Z.; Jiang, H.; Syversen, T.; Rocha, J.B.; Farina, M.; Aschner, M. The methylmercury-l-cysteine conjugate is a substrate for the l-type large neutral amino acid transporter. *J. Neurochem.* **2008**, *107*, 1083–1090. [CrossRef] [PubMed]

7.   Hirayama, K. Effect of amino acids on brain uptake of methyl mercury. *Toxicol. Appl. Pharmacol.* **1980**, *55*, 318–323. [CrossRef]

8.   Aschner, M.; Clarkson, T.W. Methyl mercury uptake across bovine brain capillary endothelial cells in vitro: The role of amino acids. *Pharmacol. Toxicol.* **1989**, *64*, 293–297. [CrossRef] [PubMed]

9.   Takeuchi, T.; Eto, K. *The Pathology of Minamata Disease*; Kyushu University Press: Fukuoka, Japan, 1999.

10. Study Group. *Minamata Disease*; Kumamoto University: Kumamoto, Japan, 1966.

11. Bakir, F.; Damluji, S.F.; Amin-Zaki, L.; Murtadha, M.; Khalidi, A.; al-Rawi, N.Y.; Tikriti, S.; Dahahir, H.I.; Clarkson, T.W.; Smith, J.C.; et al. Methylmercury poisoning in Iraq. *Science* **1973**, *181*, 230–241. [CrossRef] [PubMed]

12. Amin-Zaki, L.; Elhassani, S.; Majeed, M.A.; Clarkson, T.W.; Doherty, R.A.; Greenwood, M.R.; Giovanoli-Jakubczak, T. Perinatal methylmercury poisoning in Iraq. *Am. J. Dis. Child.* **1976**, *130*, 1070–1076. [CrossRef] [PubMed]

13. Harada, M. Congenital minamata disease: Intrauterine methylmercury poisoning. *Teratology* **1978**, *18*, 285–288. [CrossRef] [PubMed]

14. Nishigaki, S.; Harada, M. Methylmercury and selenium in umbilical cords of inhabitants of the minamata area. *Nature* **1975**, *258*, 324–325. [CrossRef] [PubMed]

15. Sakamoto, M.; Murata, K.; Tsuruta, K.; Miyamoto, K.; Akagi, H. Retrospective study on temporal and regional variations of methylmercury concentrations in preserved umbilical cords collected from inhabitants of the minamata area, Japan. *Ecotoxicol. Environ. Saf.* **2010**, *73*, 1144–1149. [CrossRef] [PubMed]

16. Sakamoto, M.; Nakano, A.; Akagi, H. Declining minamata male birth ratio associated with increased male fetal death due to heavy methylmercury pollution. *Environ. Res.* **2001**, *87*, 92–98. [CrossRef] [PubMed]

17. Harada, M.; Akagi, H.; Tsuda, T.; Kizaki, T.; Ohno, H. Methylmercury level in umbilical cords from patients with congenital minamata disease. *Sci. Total Environ.* **1999**, *234*, 59–62. [CrossRef]

18. Akagi, H.; Grandjean, P.; Takizawa, Y.; Weihe, P. Methylmercury dose estimation from umbilical cord concentrations in patients with minamata disease. *Environ. Res.* **1998**, *77*, 98–103. [CrossRef] [PubMed]

19. Sakamoto, M.; Murata, K.; Domingo, J.L.; Yamamoto, M.; Oliveira, R.B.; Kawakami, S.; Nakamura, M. Implications of mercury concentrations in umbilical cord tissue in relation to maternal hair segments as biomarkers for prenatal exposure to methylmercury. *Environ. Res.* **2016**, *149*, 282–287. [CrossRef] [PubMed]

20. Mocarelli, P.; Gerthoux, P.M.; Ferrari, E.; Patterson, D.G., Jr.; Kieszak, S.M.; Brambilla, P.; Vincoli, N.; Signorini, S.; Tramacere, P.; Carreri, V.; et al. Paternal concentrations of dioxin and sex ratio of offspring. *Lancet* **2000**, *355*, 1858–1863. [CrossRef]

21. Tokuomi, H.; Okajima, T.; Kanai, J.; Tsunoda, M.; Ichiyasu, Y.; Misumi, H.; Shimomura, K.; Takaba, M. Minamata disease. *World Neurol.* **1961**, *2*, 536–545. [PubMed]

22. Wakabayashi, K.; Kakita, A.; Sakamoto, M.; Su, M.; Iwanaga, K.; Ikuta, F. Variability of brain lesions in rats administered methylmercury at various postnatal development phases. *Brain Res.* **1995**, *705*, 267–272. [CrossRef]

23. Sakamoto, M.; Wakabayashi, K.; Kakita, A.; Hitoshi, T.; Adachi, T.; Nakano, A. Widespread neuronal degeneration in rats following oral administration of methylmercury during the postnatal developing phase: A model of fetal-type minamata disease. *Brain Res.* **1998**, *784*, 351–354. [CrossRef]

24. Aschner, M.; Clarkson, T.W. Mercury 203 distribution in pregnant and nonpregnant rats following systemic infusions with thiol-containing amino acids. *Teratology* **1987**, *36*, 321–328. [CrossRef] [PubMed]

25. Kajiwara, Y.; Yasutake, A.; Adachi, T.; Hirayama, K. Methylmercury transport across the placenta via neutral amino acid carrier. *Arch. Toxicol.* **1996**, *70*, 310–314. [CrossRef] [PubMed]

26. Sakamoto, M.; Kakita, A.; Wakabayashi, K.; Takahashi, H.; Nakano, A.; Akagi, H. Evaluation of changes in methylmercury accumulation in the developing rat brain and its effects: A study with consecutive and moderate dose exposure throughout gestation and lactation periods. *Brain Res.* **2002**, *949*, 51–59. [CrossRef]

27. Honnda, S.; Hylandaer, L.; Sakamoto, M. Recent advances in evaluation of health effects on mercury with special reference to methylmercury—A minireview. *Environ. Health Prev. Med.* **2006**, *11*, 171–176. [CrossRef] [PubMed]

28. Sakamoto, M.; Kubota, M.; Liu, X.J.; Murata, K.; Nakai, K.; Satoh, H. Maternal and fetal mercury and n-3 polyunsaturated fatty acids as a risk and benefit of fish consumption to fetus. *Environ. Sci. Technol.* **2004**, *38*, 3860–3863. [CrossRef] [PubMed]

29. Sakamoto, M.; Kubota, M.; Matsumoto, S.; Nakano, A.; Akagi, H. Declining risk of methylmercury exposure to infants during lactation. *Environ. Res.* **2002**, *90*, 185–189. [CrossRef]

30. Sakamoto, M.; Yasutake, A.; Domingo, J.L.; Chan, H.M.; Kubota, M.; Murata, K. Relationships between trace element concentrations in chorionic tissue of placenta and umbilical cord tissue: Potential use as indicators for prenatal exposure. *Environ. Int.* **2013**, *60*, 106–111. [CrossRef] [PubMed]

31. Stern, A.H.; Smith, A.E. An assessment of the cord blood: Maternal blood methylmercury ratio: Implications for risk assessment. *Environ. Health Perspect.* **2003**, *111*, 1465–1470. [CrossRef] [PubMed]

32. Arbuckle, T.E.; Liang, C.L.; Morisset, A.S.; Fisher, M.; Weiler, H.; Cirtiu, C.M.; Legrand, M.; Davis, K.; Ettinger, A.S.; Fraser, W.D.; et al. Maternal and fetal exposure to cadmium, lead, manganese and mercury: The mirec study. *Chemosphere* **2016**, *163*, 270–282. [CrossRef] [PubMed]

33. Hu, Y.; Chen, L.; Wang, C.; Zhou, Y.; Zhang, Y.; Wang, Y.; Shi, R.; Gao, Y.; Tian, Y. Prenatal low-level mercury exposure and infant neurodevelopment at 12 months in rural northern china. *Environ. Sci. Pollut. Res. Int.* **2016**, *23*, 12050–12059. [CrossRef] [PubMed]

34. Song, Y.; Lee, C.K.; Kim, K.H.; Lee, J.T.; Suh, C.; Kim, S.Y.; Kim, J.H.; Son, B.C.; Kim, D.H.; Lee, S. Factors associated with total mercury concentrations in maternal blood, cord blood, and breast milk among pregnant women in busan, korea. *Asia Pac. J. Clin. Nutr.* **2016**, *25*, 340–349. [PubMed]

35. Huang, S.H.; Weng, K.P.; Lin, C.C.; Wang, C.C.; Lee, C.T.; Ger, L.P.; Wu, M.T. Maternal and umbilical cord blood levels of mercury, manganese, iron, and copper in southern Taiwan: A cross-sectional study. *J. Chin. Med. Assoc.* **2017**, *80*, 442–451. [CrossRef] [PubMed]

36. Vigeh, M.; Nishioka, E.; Ohtani, K.; Omori, Y.; Matsukawa, T.; Koda, S.; Yokoyama, K. Prenatal mercury exposure and birth weight. *Reprod. Toxicol.* **2018**, *76*, 78–83. [CrossRef] [PubMed]

37. Sakamoto, M.; Chan, H.M.; Domingo, J.L.; Koriyama, C.; Murata, K. Placental transfer and levels of mercury, selenium, vitamin e, and docosahexaenoic acid in maternal and umbilical cord blood. *Environ. Int.* **2018**, *111*, 309–315. [CrossRef] [PubMed]

38. Rice, D.; Barone, S., Jr. Critical periods of vulnerability for the developing nervous system: Evidence from humans and animal models. *Environ. Health Perspect.* **2000**, *108*, 511–533. [CrossRef] [PubMed]

39. WHO (World Health Organization). *Children's Exposure to Mercury Compounds*; World Health Organization: Geneva, Switzerland, 2010.

40. Boischio, A.A.; Cernichiari, E.; Henshel, D. Segmental hair mercury evaluation of a single family along the upper madeira basin, Brazilian Amazon. *Cad. Saude Publica* **2000**, *16*, 681–686. [CrossRef] [PubMed]

41. Cernichiari, E.; Brewer, R.; Myers, G.J.; Marsh, D.O.; Lapham, L.W.; Cox, C.; Shamlaye, C.F.; Berlin, M.; Davidson, P.W.; Clarkson, T.W. Monitoring methylmercury during pregnancy: Maternal hair predicts fetal brain exposure. *Neurotoxicology* **1995**, *16*, 705–710. [PubMed]

42. Alfthan, G.V. Toenail mercury concentration as a biomarker of methylmercury exposure. *Biomarkers* **1997**, *2*, 233–238. [CrossRef] [PubMed]

43. Hinners, T.; Tsuchiya, A.; Stern, A.H.; Burbacher, T.M.; Faustman, E.M.; Marien, K. Chronologically matched toenail-hg to hair-hg ratio: Temporal analysis within the Japanese community (U.S.). *Environ. Health* **2012**, *11*, 81. [CrossRef] [PubMed]

44. Ohno, T.; Sakamoto, M.; Kurosawa, T.; Dakeishi, M.; Iwata, T.; Murata, K. Total mercury levels in hair, toenail, and urine among women free from occupational exposure and their relations to renal tubular function. *Environ. Res.* **2007**, *103*, 191–197. [CrossRef] [PubMed]

45. Sakamoto, M.; Chan, H.M.; Domingo, J.L.; Oliveira, R.B.; Kawakami, S.; Murata, K. Significance of fingernail and toenail mercury concentrations as biomarkers for prenatal methylmercury exposure in relation to segmental hair mercury concentrations. *Environ. Res.* **2015**, *136*, 289–294. [CrossRef] [PubMed]

46. Grandjean, P.; Budtz-Jørgensen, E.; Jørgensen, P.J.; Weihe, P. Umbilical cord mercury concentration as biomarker of prenatal exposure to methylmercury. *Environ. Health Perspect.* **2005**, *113*, 905–908. [CrossRef] [PubMed]

toxics

MDPI

Article

Methylmercury Exposure and Developmental Outcomes in Tohoku Study of Child Development at 18 Months of Age

Nozomi Tatsuta [1,†]**, Kunihiko Nakai** [1,*,†]**, Mineshi Sakamoto** [1,2]**, Katsuyuki Murata** [3] **and Hiroshi Satoh** [4,*]

[1] Development and Environmental Medicine, Tohoku University Graduate School of Medicine, 2-1 Seiryo-machi, Aoba-ku, Sendai 980-8575, Japan; nozomi@med.tohoku.ac.jp (N.T.); sakamoto@nimd.go.jp (M.S.)

[2] Environmental Health Section, Department of Environmental Science and Epidemiology, National Institute for Minamata Disease, Kumamoto 867-0008, Japan

[3] Department of Environmental Health Sciences, Akita University Graduate School of Medicine, Akita 010-8502, Japan; winestem@med.akita-u.ac.jp

[4] Environmental Health Science, Tohoku University Graduate School of Medicine, Sendai 980-8575, Japan

[*] Correspondence: nakaik@med.tohoku.ac.jp (K.N.); h.satoh@med.tohoku.ac.jp (H.S.); Tel.: +81-22-717-8949 (K.N.)

[†] These authors contributed equally to this work.

Received: 29 July 2018; Accepted: 18 August 2018; Published: 21 August 2018

Abstract: Seafood is an important component in a healthy diet and may contain methylmercury or other contaminants. It is important to recognize the risks and benefits of consuming seafood. A longitudinal prospective birth cohort study has been conducted to clarify the effects of neurotoxicants on child development—the Tohoku Study of Child Development (TSCD) in Japan. TSCD comprises two cohorts; a polychlorinated biphenyls (PCB) cohort (urban area) and a methylmercury cohort (coastal area). Our previous results from the coastal area showed prenatal methylmercury exposure affected psychomotor development in 18-month-olds, and boys appear to be more vulnerable to the exposure than girls. In this report, we have added the urban area cohort and we reanalyzed the impact of prenatal exposure to methylmercury, which gave the same results as before. These findings suggest prenatal exposure to low levels methylmercury may have adverse effects on child development, especially in boys.

Keywords: methylmercury; prenatal exposure; child development

1. Introduction

Seafood is a good source of protein and various essential nutrients including n-3 polyunsaturated fatty acids (n-3 PUFAs), selenium, iodine, and vitamin D, whereas it is low in fatty acids. Among the nutrients, n-3 PUFAs, such as docosahexaenoic acid (DHA), are important because these fats are difficult to get in sufficient amounts from other food items but are highly beneficial for maternal and fetal health because of their critical role in proper brain development and function [1,2]. However, seafood also contains several toxic chemicals like methylmercury and polychlorinated biphenyls (PCB) due to bioaccumulation in the aquatic food chain [3,4]. Human exposure to methylmercury and PCB occurs mainly from the intake of seafood [5]. Indeed, mercury exposure level is related to the amount of seafood intake [6], and several studies have confirmed a relationship between seafood intake and mercury blood/hair levels in humans [6–8]. Importantly, because methylmercury readily crosses the placenta, fetuses are at a high risk to exposure [9]. Many previous studies found an association between prenatal exposure to methylmercury and adverse effects on child neurodevelopment [10–17].

Because the Japanese are one of the world's largest seafood consumers, we initiated the Tohoku Study of Child Development (TSCD) in 2001 [18]. TSCD is a prospective birth cohort study that investigates the effects of neurotoxicants on child development in Japan with the objective to determine the potential risks and benefits of seafood eating during pregnancy. TSCD consists of two birth cohorts, an urban area and a coastal area cohort, in northeastern Japan [19–24]. The purpose of the urban area cohort is to examine the effect of PCB exposure, and that of the coastal area cohort is to examine the effect of prenatal methylmercury exposure on child development. The chemical substances being measured are different, but the protocols are almost the same. We previously reported that the psychomotor development in 18-month-old children of the coastal area was significantly correlated with cord-blood total mercury (THg) only in the boys, and that this association remained significant after adjustment for possible confounders, including maternal-plasma DHA and cord-plasma selenium [22]. This result suggests that boys are more vulnerable to methylmercury exposure than girls, which is consistent with a few other studies [22,23,25–27]; although most other studies on this subject did not include gender-specific analyses [28]. Even if the gender was considered, most birth cohort studies in children included it as a confounding variable, which needed to be controlled and generalized the chemotoxic effects without accounting for possible gender discrepancies [29].

In this report, we introduce the exposure status of methylmercury in Japanese pregnant women, and the protocol of TSCD. Then, we also measured methylmercury exposure levels in an urban area cohort, therefore, we investigated the impact of prenatal methylmercury exposure on the developmental scores using the Bayley scales of infant development second edition (BSID-II) [30] in the urban and coastal areas for 18-month-old children, with an emphasis on the potential impact of the child gender.

2. Materials and Methods

2.1. TSCD Outline

The TSCD protocols have been described in a previous report [18]. TSCD comprises two birth cohorts from northeastern Japan, an urban area and a coastal area cohort. The first birth cohort study had been conducted in an urban area to examine the effects of perinatal low-level PCB exposure on child developmental outcomes. The second birth cohort study is being conducted to examine the effects of prenatal methylmercury exposure to child development outcomes. To decide the research field, we investigated an area with a high methylmercury exposure level. We conducted a preliminary survey to select the research area from four candidate areas in the Tohoku region. To identify the area with the highest hair THg concentration, we recruited women who visited their regional obstetrical-gynecological clinics to provide their hair samples for THg analysis. Written informed consent was obtained from each participant prior to collecting the hair samples. In addition, we referred to a study by Yasutake et al. [31], who reported hair THg concentration data from five districts in Japan. Based on all these results, we had finally decided to conduct the coastal area for our research (Candidate area A of Table 1).

To establish an optimal study population, the eligibility criteria included a singleton pregnancy, Japanese as the mother tongue, and neonates born at term (36–42 weeks of gestation) with a birth weight of more than 2400 g and no congenital anomalies or diseases. We referred to the Dutch PCB/Dioxin study before starting our cohort study [32]. One of the inclusion criteria of the Dutch study is child born at term (37–42 weeks). This is because the reason for preterm birth is thought to be affected by factors other than prenatal PCB exposure. Therefore, in our study, low birth weight children and preterm birth children were excluded. Outcome measurements of age-appropriate neurobehavioral assessments and parent report questionnaires were selected to examine the effect of prenatal methylmercury exposure. The details of the follow-up neurobehavioral assessments and questionnaires are shown in Table 2.

The study protocol was approved by the Medical Ethics Committee of the Tohoku University Graduate School of Medicine.

Table 1. Geometric mean and min/max of hair THg * in candidate areas of our research area (µg/g).

Candidate Area	n	Geometric Mean	Min	Max
Candidate area A	100	3.27	1.00	13.3
Candidate area B	94	1.99	0.66	10.3
Candidate area C	99	1.80	0.55	5.35
Candidate area D	100	2.01	0.67	8.15
Yasutake et al. *Tohoku J. Exp. Med.* **2003**, *199(3)*, 161–169				
Minamata	594	1.23	0.09	7.33
Kumamoto	327	1.33	0.14	6.20
Tottori	209	1.40	0.26	12.5
Wakayama	303	1.40	0.00	8.09
Chiba	233	2.30	0.14	25.8

* Total mercury.

Table 2. Follow-up and outcome measures of TSCD *.

Child Age	Neurobehavioral Development Assessment
3 days	Neonatal Behavioral Assessment Scale
7 months	Kyoto Scale of Psychological Development (KSPD)
	Bayley Scales of Infant Development second edition (BSID-II)
	Fagan Test of Infant Intelligence
18 months (1.5 years)	KSPD, BSID-II, Evaluation of Environmental Stimulation (EES)
	Raven standard progressive matrices
30 months (2.5 years)	Child Behavior Checklist age for 2–3, EES
42 months (3.5 years)	Kaufman Assessment Battery for Children
66 months (5.5 years)	Social-Maturity Skill Scale (S-M scale)
84 months (7 years)	Wechsler Intelligence Scale for Children Third edition
120 months (10 years)	S-M scale
144 months (12 years)	Wechsler Intelligence Scale for Children Forth edition

* Tohoku Study of Child Development.

2.1.1. Urban Area Cohort (PCB Cohort)

For the urban area cohort, we recruited 1500 of 22-week-pregnant women from January 2001 to September 2003. Of these, 687 pregnant women agreed to participate (participation rate, 45.8%) and provided written informed consent to provisionally register for this study. After birth, 88 provisional registrants withdrew from the study due to various reasons in accordance with the eligibility criteria, for example, premature birth, and being transferred to another hospital. A total of 599 mother–child pairs were finally registered. The flowchart of how the study participants were determined is shown in Figure 1a.

2.1.2. Coastal Area Cohort (Methylmercury Cohort)

For the coastal area cohort, we recruited 1312 of 22-week-pregnant women from December 2002 to March 2006; 879 pregnant women agreed to participate (participation rate, 67.0%) and provided written informed consent to provisionally register for this study. While the initial registration procedure was the same as that for the urban area cohort, a screening step for identifying participants with high hair THg levels was added. Immediately after the recruitment, hair samples were obtained and analyzed for THg content. Potential participants with a high THg level were registered together with a matching number of participants with a low THg level. These low THg participants were selected with adjusting the age, sex, amount of seafood consumption, and the tendency to consume certain fish species. At this step, 102 participants were excluded from the study, because participants had lower hair THg. Finally, 749 mother–child pairs were registered. The flowchart of how the study participants were determined is shown in Figure 1b.

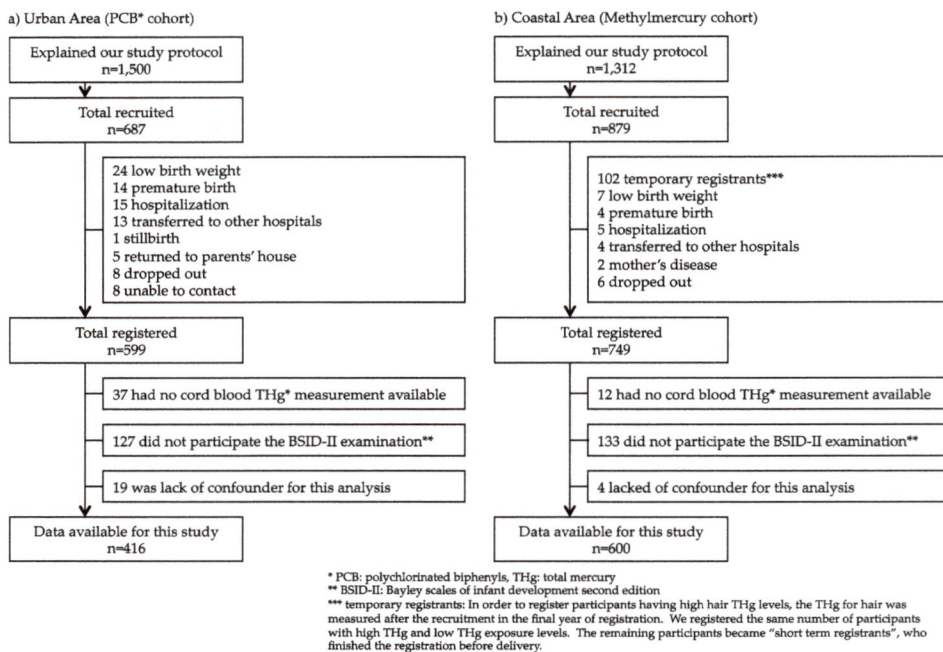

a) Urban Area (PCB* cohort)

Explained our study protocol
n=1,500

Total recruited
n=687

24 low birth weight
14 premature birth
15 hospitalization
13 transferred to other hospitals
1 stillbirth
5 returned to parents' house
8 dropped out
8 unable to contact

Total registered
n=599

37 had no cord blood THg* measurement available

127 did not participate the BSID-II examination**

19 was lack of confounder for this analysis

Data available for this study
n=416

b) Coastal Area (Methylmercury cohort)

Explained our study protocol
n=1,312

Total recruited
n=879

102 temporary registrants***
7 low birth weight
4 premature birth
5 hospitalization
4 transferred to other hospitals
2 mother's disease
6 dropped out

Total registered
n=749

12 had no cord blood THg* measurement available

133 did not participate the BSID-II examination**

4 lacked of confounder for this analysis

Data available for this study
n=600

* PCB: polychlorinated biphenyls, THg: total mercury
** BSID-II: Bayley scales of infant development second edition
*** temporary registrants: In order to register participants having high hair THg levels, the THg for hair was measured after the recruitment in the final year of registration. We registered the same number of participants with high THg and low THg exposure levels. The remaining participants became "short term registrants", who finished the registration before delivery.

Figure 1. Flowchart of study population. Flowchart indicating the number of mother–child pairs included and excluded from the study population.

2.2. Exposure Markers

THg concentrations in whole cord blood, hair, and breast milk were analyzed using cold vapor atomic absorption spectrometry [19,20]. The analytical method for THg has been described elsewhere [20,33]. Cord blood and breast milk PCB were analyzed by high-resolution gas chromatography and high-resolution mass spectrometry using the isotope dilution method. Laboratory analytical methods and quality control procedures were described elsewhere [19,34]. Cord blood lead concentrations in whole cord blood were determined by inductively coupled plasma mass spectrometry (ICP-MS, SRL Inc., Tokyo, Japan) [19,22,23]. Cord blood selenium was analyzed by ICP-MS, and cord plasma selenium was analyzed by Watkinson method [22,23]. Maternal plasma DHA was analyzed by using gas chromatography [22].

2.3. Outcome

The BSID-II score was one of the study outcomes used for 18-month-old children. The raw scores were standardized for the child's age in days at the time of test administration. The raw scores of each scale were converted into the mental developmental index (MDI) and psychomotor developmental index (PDI), based on age-appropriate norms. These scores are derived from the total raw scores of each test and normalized on a score scale with a mean of 100 and a standard deviation (SD) of 15. Since there was no standardized version of the BSID-II for Japan, we prepared a Japanese version ourselves. The reliability of administration of the BSID-II has been described elsewhere [35].

2.4. Confounding Variables

Information about pregnancy, delivery, and the characteristics of infants at birth such as child gender, birth weight, and birth order were obtained from medical records. We obtained information about demographics and smoking/drinking habits during pregnancy (presence/absence) from a questionnaire four days after delivery.

Maternal seafood intake during pregnancy was assessed using a food frequency questionnaire (FFQ) that was administered by trained interviewers immediately after delivery. Methylmercury intake was estimated from seafood intake and methylmercury concentrations in each type of seafood. The calculation method has been described elsewhere [36].

The maternal intellectual ability was evaluated using the Raven standard progressive matrices [37]. We used the raw score for analysis because it has not been standardized in Japan. Home environment was assessed using the Evaluation of Environmental Stimulation (EES) questionnaire [38], which had been established in Japan modified after home observation for measurement of the environment score [39]. The mother was asked to fill out the Raven standard progressive matrices and the EES when her child was 18 months of age.

2.5. Statistics

The THg in cord blood and maternal hair were logarithmically transformed ($\log_{10}$) because of skewed distributions. Sex differences in basal characteristics, exposure levels, and scores of the BSID-II were analyzed using the Student *t*-test, Mann–Whitney *U* test, or Fisher exact test. Multiple regression analysis was used to adjust for possible confounders. Independent variables in the analysis were child gender, birth weight, birth order, drinking and smoking habits, the Raven scores, EES score, and testers of the BSID-II (for which dummy variables were used). All analyses using two-sided *p*-values, were performed using SPSS Ver. 23.0 (SPSS Inc., Tokyo, Japan, 2016) and the statistical significance was set at $p < 0.05$.

3. Results

In the urban area cohort (PCB cohort) with 599 registered mothers, cord blood samples were collected from 562 participants (93.8%), maternal hair samples at parturition were collected from 595 participants (99.3%), and the FFQ was administered to 598 participants (99.9%). In the coastal area cohort (methylmercury cohort) with 749 registered mothers, cord blood samples were collected from 731 participants (97.6%), maternal hair samples at parturition were collected from 748 participants (99.9%), and the FFQ was administered to all participants (100.0%). In Japan, the tolerable weekly intake (TWI) for methylmercury of 2.0 µg/kg body weight per week for pregnant and potentially pregnant women was proposed by the Japan Food Safety Commission [40]. However, 12.4% of the urban area participants and 18.2% of the coastal area participants exceeded the TWI [22,36].

In this study, we excluded children who did not have cord blood samples for methylmercury biomarker, did not participate in the 18-months BSID-II examination, or lacked a confounder for this analysis. Thus, for the final analysis, 416 mother–child pairs from the urban area and 600 mother–child pairs from coastal area were included (Figure 1). The mean age of the participating children was 18 months (range, 17 to 24 months). Table 3 shows the basal characteristics of the mother–child pairs. By comparing the urban and coastal area cohorts, differences were found in many variables, such as maternal age at parturition, BMI before pregnancy, and drinking/smoking habits during pregnancy. The distributions of the biomarkers are shown in Table 4. Although the methylmercury biomarkers were significantly higher in coastal area participants than in urban ones, there was no difference in total seafood intake during pregnancy. Since the amount and species of seafood consumed were expected to be different among the districts in Japan, we examined the differences in intake of each kind of seafood (Table 5). Coastal area participants consumed more bonito, salmon, shellfish, some other kind of fish, and shellfish than the urban area participants, who consumed more whale, yellowtail, and eel. The type of seafood eaten by coastal area participants had high concentrations of methylmercury such as bonito.

Table 3. Background characteristics of TSCD * mother–child pairs.

Basal Characterisctics	Urban Area Mean ± SD * (or %)	Coastal Area Mean ± SD * (or %)	*p*-Value **
Maternal characteristics			
Maternal age at parturition (years)	31.3 ± 4.4	29.5 ± 4.9	$p < 0.001$
Body mass index before pregnancy (kg/m^2)	21.0 ± 2.8	21.5 ± 3.3	0.002
Drinkers during pregnancy (%, yes)	31.7	16.6	$p < 0.001$
Smokers during pregnancy (%, yes)	7.8	12.5	$p < 0.001$
Maternal education level (%, >12 y)	74.9	41.5	$p < 0.001$
Raven score ***	51.5 ± 6.3	49.9 ± 6.0	$p < 0.001$
EES score at 18 months ***	28.2 ± 3.4	26.6 ± 3.8	$p < 0.001$
Child characteristics			
Child gender (%, boys)	52.6	50.9	0.547
Birth order (%, first child)	51.4	42.1	0.001
Gestational duration (weeks)	39.5 ± 1.3	39.7 ± 1.2	0.038
Birth weight (g)	3073 ± 338	3141 ± 365	$p < 0.001$
Delivery type (%, vaginal delivery)	83.6	84.4	0.765
Apgar score (1 min)	8.2 ± 0.8	8.4 ± 0.8	$p < 0.001$

* TSCD, Tohoku Study of Child Development; SD, standard deviation. ** Student *t*-test or χ^2 test. *** Raven, Raven standard progressive matrices; EES, Evaluation of Environmental Stimulation.

Table 4. Exposures indices in urban and coastal area participants.

Exposures	Urban Area n, Median, 5–95 Percentiles	Coastal Area n, Median, 5–95 Percentiles	*p*-Value *
Exposure biomarkers:			
Cord-blood THg (ng/g) **	562, 10.0, 4.2–22.4	731, 16.0, 5.6–39.3	$p < 0.001$
Maternal hair THg (µg/g) **	595, 2.0, 0.9–4.4	748, 2.6, 0.9–6.0	$p < 0.001$
Breast milk THg (ng/g) **	-	27, 0.8, 0.1–1.8	-
Cord-blood PCB (ng/g-lipid) **	518, 45.8, 18.4–112.2	-	-
Breast milk PCB (ng/g-lipid) **	544, 93.1, 42.4–185.9	-	-
Cord-blood lead (ng/dL)	555, 1.0, 0.6–1.8	664, 0.7, 0.4–1.4	$p < 0.001$
Cord-blood selenium (ng/mL)	555, 192.7 (130.3–271.9)	-	-
Cord-plasma selenium (ng/g)	-	709, 66.3, 51.0–271.9	-
Maternal-plasma DHA **	-	742, 169.7, 101.1–256.9	-
Seafood intake during pregnancy (kg/y)	598, 44.4, 12.6–110.8	749, 47.7, 10.5–140.6	0.089

* Mann–Whitney *U* test. ** THg, total mercury; PCB, polychlorinated biphenyls; DHA, docosahexaenoic acid.

The participation rate of each examination is shown in Table 6. The urban area cohort was closed after the examination at the age of 84 months due to the lack of research funds. However, while we were examining the 84-month-old children of the coastal area cohort, the Great East Japan Earthquake hit, causing severe damage. Because of this disaster, the participants were compulsorily divided into predisaster and postdisaster groups [21]. The follow-up rate for the 84-months-old examination of the predisaster group was 78.1%, but for the postdisaster group, it was 65.6%. Therefore, the follow-up rate for the 84-months-old examination was lower than the other examination.

Table 5. Comparison of the amount of intake of seafood determined by the FFQ * (g/day).

| Fish Species | Urban Area (*n* = 598) | Coastal Area (*n* = 749) | *p*-Value ** |
	Median (Min–Max)	Median (Min–Max)	
Tuna	4.1 (0.0–123.7)	4.4 (0.0–105.0)	0.161
Bonito	2.1 (0.0–41.8)	2.7 (0.0–108.3)	*p* < 0.001
Whale	0.0 (0.0–17.5)	0.0 (0.0–2.3)	0.004
Salmon	3.1 (0.0–34.7)	3.1 (0.0–92.5)	0.037
Eel	0.3 (0.0–32.1)	0.0 (0.0–12.5)	*p* < 0.001
Yellowtail	0.6 (0.0–70.0)	0.0) (0.0–70.0)	*p* < 0.001
Silvery blue fish	5.8 (0.0–55.0)	5.8 (0.0–70.0)	0.546
White-meat fish	7.2 (0.0–87.0)	7.2 (0.0–87.0)	0.918
Other fish	0.0 (0.0–57.9)	3.0 (0.0–90.0)	*p* < 0.001
Squid/Octopus	2.0 (0.0–30.0)	2.0 (0.0–60.0)	0.264
Shellfish	1.7 (0.0–39.3)	2.3 (0.0–50.0)	*p* < 0.001
Salmon roe	0.0 (0.0–37.5)	0.0 (0.0–37.5)	0.162
Canned tuna	1.7 (0.0–60.0)	2.0 (0.0–60.0)	0.524

* FFQ, food frequency questionnaire. ** Mann–Whitney *U* test.

Table 6. Follow-up rates of each examination.

| Time of Each Examination | Urban Area | | | Coastal Area | | |
	Registrants	Participants	%	Registrants	Participants	%
3 days	599	587	98.0	749	709	94.7
7 months	594	516	86.9	749	653	87.2
18 months (1.5 years)	589	477	81.0	747	617	82.6
30 months (2.5 years)	595	499	83.9	739	649	87.8
42 months (3.5 years)	566	472	83.4	733	597	81.3
66 months (5.5 years)	580	456	78.6	727	614	84.5
84 months (7 years)	546	457	83.7	720	498	69.2
120 months (10 years)				711	569	80.0
144 months (12 years)				699	385	55.0

The scores of the BSID-II are provided in Table 7. The MDI of the BSID-II was significantly higher in the urban area children than that in the coastal area children, whereas there was no difference in the PDI of the BSID-II between these two groups. Table 8 shows the results of the multiple regression analysis. Cord-blood THg was not significantly correlated with any BSID-II scores. However, there was a relationship between child gender and BSID-II scores. Therefore, the following analysis was carried out separately for boys and girls. As shown in Table 9, it was only in boys that the cord-blood THg was significantly associated with lower PDI of the BSID-II. As we focused on the PDI score of BSID-II and stratified the participants by research area, the association between cord-blood THg and PDI was found only in the coastal area boy group (Table 10).

Table 7. Scores of the BSID-II (Mean ± SD *) at 18 months of age (*n* = 1016).

| BSID-II Scores | Urban Area (*n* = 416) | Coastal Area (*n* = 600) | *p*-Value ** |
	Mean ± SD *	Mean ± SD *	
MDI ***	89.8 ± 11.9	86.9 ± 10.6	<0.001
PDI ***	84.6 ± 10.6	84.4 ± 10.6	0.793

* SD, standard deviation. ** Student *t*-test. *** MDI, mental developmental index; PDI, psychomotor developmental index.

Table 8. Relations of cord-blood THg * and possible confounders to the scores of BSID-II: Standardized regression coefficients (β) of multiple regression analysis (n = 1016).

Major Independent Variables	MDI **		PDI **	
	β	p-Value	β	p-Value
Cord-blood THg *	−0.028	0.380	−0.053	0.104
Child gender	−0.230	<0.001	−0.111	<0.001
Birth weight	0.034	0.261	−0.012	0.696
Birth order	−0.031	0.312	0.049	0.113
Drinking habit during pregnancy	0.035	0.253	−0.044	0.164
Smoking habit during pregnancy	0.000	0.997	0.040	0.197
Raven score ***	0.036	0.238	0.063	0.044
EES score at 18 months ***	0.134	<0.001	0.071	0.025
Contribution rate, R^2	0.101	<0.001	0.080	<0.001

Other independent variables: Testers of the BSID-II and research area. Cord-blood THg was logarithmically transformed. * THg, total mercury. ** MDI, mental developmental index; PDI, psychomotor developmental index. *** Raven, Raven standard progressive matrices; EES, Evaluation of Environmental Stimulation.

Table 9. Relations of cord-blood THg ** and possible confounders to the scores of BSID-II: Standardized regression coefficients (β) of multiple regression analysis.

Major Independent Variables	Boys (n = 523)				Girls (n = 493)			
	MDI *		PDI *		MDI *		PDI *	
	β	p-Value	β	p-Value	β	p-Value	β	p-Value
Cord-blood THg **	−0.036	0.437	−0.122	0.008	−0.017	0.729	0.024	0.616
Birth weight	0.093	0.036	0.045	0.307	−0.051	0.262	−0.085	0.057
Birth order	0.007	0.873	0.026	0.554	−0.085	0.061	0.066	0.142
Drinking habit during pregnancy	−0.021	0.639	−0.039	0.379	0.085	0.062	−0.053	0.236
Smoking habit during pregnancy	0.000	0.999	0.033	0.440	0.012	0.782	0.057	0.203
Raven score ***	0.052	0.239	0.036	0.407	0.006	0.888	0.090	0.045
EES score at 18 months ***	0.092	0.038	0.086	0.052	0.204	<0.001	0.068	0.137
Contribution rate, R^2	0.058	<0.001	0.078	<0.001	0.061	<0.001	0.082	<0.001

Other independent variables: testers of the BSID-II and research area. Cord-blood THg was logarithmically transformed. * MDI, mental developmental index; PDI, psychomotor developmental index. ** THg, total mercury. *** Raven, Raven standard progressive matrices; EES, Evaluation of Environmental Stimulation.

Table 10. Relations of cord-blood THg * and possible confounders to the PDI ** of BSID-II: Standardized regression coefficients (β) of multiple regression analysis only in boys.

Major Independent Variables	Urban Area (n = 220)		Coastal Area (n = 303)	
	β	p-Value	β	p-Value
Cord-blood THg *	−0.033	0.606	−0.18	0.002
Birth weight	−0.045	0.477	0.091	0.128
Birth order	−0.081	0.202	0.091	0.124
Drinking habit during pregnancy	−0.058	0.362	−0.044	0.450
Smoking habit during pregnancy	−0.016	0.806	0.062	0.282
Raven score ***	−0.098	0.126	0.124	0.033
EES score at 18 months ***	0.039	0.543	0.116	0.048
Contribution rate, R^2	0.148	<0.001	0.052	0.007

Other independent variables: testers of the BSID-II and research area. Cord-blood THg was logarithmically transformed. * THg, total mercury. ** PDI, psychomotor developmental index. *** Raven, Raven standard progressive matrices; EES, Evaluation of Environmental Stimulation.

4. Discussion

4.1. Outline of the TSCD

In this report, we summarized the protocol and state of progress of the TSCD. The TSCD consists of participants in an urban area and coastal area, and both areas are in northeastern Japan; various differences were found in the basal characteristics as in Table 3. When clarifying the health effects of prenatal methylmercury exposure, the importance of collecting information of basal characteristics and confounding factors was shown. Although there was no difference in fish intake between the two cohorts, hair/cord-blood THg was significantly higher in the coastal participants. For this reason, it was shown that the coastal participants consume fish containing higher mercury levels than urban participants. Most fish samples contain methylmercury, but the concentrations vary greatly according to the fish species. In fact, the coastal participants consume seafood with higher methylmercury concentration such as bonito (Table 5).

In our cohort, follow-up rates of the coastal area have decreased to 69.2% in 84-month-old examination and 55.0% in 144-month-old examination, respectively as in Table 6. A possible reason why the follow-up rate of 84-month-old examination was decreased in the coastal area would be the effects of the Great East Japan Earthquake. The disaster also had several significant impacts on our participants and cohort. For instance, hair THg of the children was decreased by approximately 30% after the disaster. In these areas, the consumption of seafood was decreased after the disaster because of destructive damage to the fishery [24]. The disaster induced subtle deficits in verbal intelligence quotient (IQ) of 84-month-old children, probably due to a temporal blank in their education [21]. However, the follow-up rate improved at next examination at 120-month-old. In the 144-month-old examination follow-up rate was down to 55% as in Table 6. Reasons for not being able to participate in the examination include children being busy with study and sports club activities, and parents working, so their schedules do not match. Other than that, we thought that we were able to maintain a function as a cohort.

4.2. Exposure Levels

The median THg levels of maternal hair at parturition were 2.0 µg/g and 2.5 µg/g in the urban and coastal areas, respectively, which were lower than the levels reported in other studies in the Faroe Islands [41,42], the Seychelles [43], Canada [44], Brazil [6,45], and France [46], but higher than those in Italy [47], the Philippines [48], China [49], Austria [50], and the United States (US) [51]. These suggest that the level of exposure to methylmercury in our study participants was definitely not high compared with other studies. However, 93.6% of the urban area participants and 94.5% of the coastal area participants of our cohort had hair THg concentrations higher than the US Environmental Protection Agency (EPA) recommended levels (1.0 µg/g for hair THg) for optimal health. Yasutake et al. [31] reported that it would not be adequate to employ the reference dose (RfD) of the US EPA in Japan. We think that it is necessary to examine the effect of prenatal methylmercury on Japanese.

As described in the Results section, the Japan Food Safety Commission proposed TWI for THg at 2.0 µg/BW-kg/week in 2005. To apply the TWI for THg, 12.4% of pregnant women in the urban area and 18.2% of pregnant women in the coastal area exceeded it [22,36]. In Japan, the Ministry of Health, Labour, and Welfare advises that pregnant women limit consumption of certain species. However, according to the results of another of our studies, 25.5% (14/55) of women who have never been pregnant did not know about this advice, and even 47.7% (336/705) of women who had been pregnant did not know about it.

Since fish are rich in n-3 PUFAs, fish consumption is the primary source of n-3 PUFAs intake, and Saito et al. [52] reported the determinants of n-3 PUFAs status in maternal and cord blood were examined, and maternal seafood consumption was a potent factor. To avoid the adverse health effects of prenatal methylmercury while retaining the benefits provided by fish consumption, it is important to select suitable fish species and to pay attention to the amount of fish intake. For that purpose, the information and knowledge about seafood intake, including advice from the Ministry of Health, Labour, and Welfare, must be known thoroughly by pregnant women and women who may become pregnant. It is necessary to transmit correct information to next generation who become parents.

4.3. Gender-Specific Analyses

Compared to studies on other heavy metals (cadmium, manganese, and arsenic), studies on methylmercury and lead that were analyzed by based on child gender are more abundant. However, child gender-specific susceptibility to prenatal methylmercury exposure has not received sufficient attention [28]. When we look at just prenatal exposure to methylmercury, boys seem more susceptible to adverse effects than girls. Indeed, studies of human prenatal methylmercury poisoning in Iraq found that male newborns suffered worse complications from methylmercury exposure than female newborns [53]. In Minamata, Japan, it was reported that the male/female birth ratio decreased in the overall city population as well as in the fishing villages during 1955–1959, when the pollution degree of methylmercury was very high. Additionally, a decrease in maternal Minamata disease patients was observed, but this was due to the high number of stillborn male fetuses [26]. The adverse effects of methylmercury on child neurodevelopment were reported by cohort studies in northern Quebec [54], Seychelles [25,55], Faroe Islands [56,57], Massachusetts [58], Guiana [47], and the city of Zhoushan [50]. On the other hand, two studies reported the negative impact on girls [59,60]. According to our previous studies, the birth weight and psychomotor development of the boys were affected by cord-blood THg level [23]. The mechanism underlying gender differences in exposure related neurotoxicity is unknown, and information regarding gender differences in susceptibility of methylmercury is still too limited to draw any definite conclusions.

Understanding gender differences is useful for elucidating the pathways from exposure to manifestation, and may also provide new insights into prevention strategies [28]. In order to clarify this, further studies are required to clarify the mechanism of effects of prenatal exposure on gender difference in experimental studies. Also, sample size is an important factor that must be considered carefully. There are several large prospective birth cohort studies on environmental contaminants and child health, such as the Danish National Birth Cohort in Denmark [61], the Norwegian Mother and Child Cohort Study in Norway [62], the Newborns and Genotoxic exposure risks project in EU [63], and the Japan Environment and Children's Study in Japan [64]. Each of these studies is made up of more than 100,000 parent–child pairs. Such large cohorts may solve this issue in the near future.

4.4. Regional Difference

In the current study, we found that increasing cord blood THg was associated with lower PDI of the BSID-II at 18 months of age in the coastal area boys but not in the urban area boys. We can postulate several reasons for this discrepancy. One of these reasons might be that the coastal area pregnant women in our study had a higher and wider range of cord-blood and hair THg than the urban area pregnant women. The maternal hair THg at parturition in the coastal area participants of this study ranged from 0.31 µg/g to 11.0 µg/g and in the urban area participants from 0.29 µg/g to 9.34 µg/g, respectively. The extremely low and narrow range of methylmercury concentration in the urban area participants of the present study may be the reason why we obtained insignificant results. Exposure level and range of the study population may be important for detecting a significant result.

As confounders for the BSID-II scores, birth weight, score of the maternal Raven standard progressive matrices, and score of the EES were chosen in multiple regression analysis (Tables 8–10), and there were significant differences in them between the urban area cohort and the coastal area cohort (Table 3). It is common knowledge that these variables are closely related to child development. For instance, a consistent association between birth weight and child development has been established [65]. The Port Pirie Cohort Study reported that the higher the occupational prestige and maternal IQ and the better the home environment, the higher the children's cognitive function [66]. To clarify the effect of prenatal methylmercury exposure, these crucial confounders should be collected properly in future studies.

4.5. Future Vision

The Faroe cohort study showed that prenatal methylmercury related neuropsychological dysfunctions were most pronounced in the domains of language, attention, and memory, and to a lesser extent in visuospatial and motor functions at 7 and 14 years [10,40,67]. After that, they examined whether negative associations are still detectable eight years later, at age 22. The results at age 22 suggest that cognitive deficits associated with prenatal methylmercury exposure remain through young adulthood. They concluded that prenatal exposure to this marine contaminant appears to cause permanent adverse effects on cognition [68]. At the present stage, even in our cohort, the negative effect of prenatal methylmercury exposure has been observed, follow-up data are needed to determine if adverse effects occur at older ages and if such effects are determined to be related to methylmercury. We have collected data on the development at different age of our children so we are planning to analyze the effect of prenatal exposure to methylmercury on child development. As our participants are turning 12-year-old, 10 years later, we would like to reconfirm Faroe study's results in Japan.

We are also measuring toxic chemicals other than THg, for instance PCB and lead in cord-blood and blood as in Table 4, due to estimation of the combined risks of the mixture effects is required. Exposure to a mixture of chemicals is ubiquitous in real life and all children are exposed to multiple toxic chemicals [69]. Recently, gradually the number of such studies is increasing [19,70–74], however, the results pattern was not so clear. A research model with concomitant exposures is necessary for evaluating subtle effects on child development. In addition, n-3 PUFAs is essential for normal brain development. Three studies reported that the negative association between prenatal methylmercury exposure and cognitive development became apparent only after adjustment with n-3 PUFAs intake [75–77]. These studies indicated the essential nutrients such as n-3 PUFAs masked the effects of methylmercury. Future studies need to indicate the exposure levels of both beneficial nutrients and toxic substances such as methylmercury and PCB. Moreover, we are planning to identify high-risk groups with genetic susceptibilities to methylmercury.

5. Conclusions

The TSCD examined both the potential risks and benefits of eating fish during pregnancy. Our participants are turning 12-year-old, and till now, we thought that we were able to maintain a function as a cohort. Results of the TSCD suggest that even relatively low levels of exposure to methylmercury may have adverse effects on child development especially in boys. Thus, a long-term follow-up study should be conducted to obtain further information about the effects of methylmercury exposure on child development.

In Japan, the TWI for methylmercury of 2.0 μg/kg body weight per week for pregnant and potentially pregnant women was decided by the Japan Food Safety Commission (2005). 12.4% of the urban area participants and 18.2% of the coastal area participants exceeded the TWI [22,36]. However, a lot of Japanese people do not know about the TWI. Seafood is an important part of a healthy diet and contains good nutritional properties for pregnant women and fetuses. Therefore, we need to present correct information about how to eat seafood. It is necessary to transmit correct information to next generation who become parents. We want to clarify an information providing method of more efficiently and effectively providing required information to pregnant women to protect our children's lives and futures.

Author Contributions: All authors contributed to the preparation of the paper. N.T. collected the dates and critically prepared this paper; K.N. organized the cohort and carried out biological monitoring; M.S. contributed to designing and implementing the study; K.M. participated in data collection and analysis; H.S. is the ex-primary investigator, conceived the study idea, and designed the study. All the authors read and accepted the final version of the manuscript.

Funding: This study was supported partly by the Japan Ministry of the Environment. The findings and conclusions of this article are solely the responsibility of the authors and do not represent the official views of the above agency.

Acknowledgments: The authors would like to thank all participants in the study, researchers, and members of the TSCD. The findings and conclusions of this article are solely the responsibility of the authors and do not represent the official views of the above government. This research was funded by the Ministry of the Environment, Japan.

Conflicts of Interest: The authors declare no conflicts of interest.

References

1. Strain, J.J.; Yeates, A.J.; van Wijngaarden, E.; Thurston, S.W.; Mulhern, M.S.; McSorley, E.M.; Watson, G.E.; Love, T.M.; Smith, T.H.; Yost, K.; et al. Prenatal exposure to methyl mercury from fish consumption and polyunsaturated fatty acids: Associations with child development at 20 mo of age in an observational study in the Republic of Seychelles. *Am. J. Clin. Nutr.* **2015**, *101*, 530–537. [CrossRef] [PubMed]
2. Haggarty, P. Fatty acid supply to the human fetus. *Annu. Rev. Nutr.* **2010**, *30*, 237–255. [CrossRef] [PubMed]
3. Boucher, O.; Muckle, G.; Jacobson, J.L.; Carter, R.C.; Kaplan-Estrin, M.; Ayotte, P.; Dewailly, E.; Jacobson, S.W. Domain-specific effects of prenatal exposure to PCBs, mercury, and lead on infant cognition: Results from the Environmental Contaminants and Child Development Study of Nunavik. *Environ. Health Perspect.* **2014**, *122*, 310–316. [CrossRef] [PubMed]
4. Julvez, J.; Mendez, M.; Femandez-Barres, S.; Romaguera, D.; Vioque, J.; Liop, S.; Ibarluzea, J.; Cuxens, M.; Avella-Garcia, C.; Tordon, A.; et al. Maternal consumption of seafood in pregnancy and child neuropsychological development: A longitudinal study based on a population with high consumption levels. *Am. J. Epidemiol.* **2016**, *183*, 169–182. [CrossRef] [PubMed]
5. Weihe, P.; Grandjean, P.; Debes, F.; White, R. Health implications for Faroe islanders of heavy metals and PCBs from pilot whales. *Sci. Total Environ.* **1996**, *186*, 141–148. [CrossRef]
6. Marques, R.C.; Doreac, J.G.; Bastosa, R.W.; de Freitas Rebelob, M.; de Freitas Fonsecab, M.; Malm, O. Maternal mercury exposure and neuro-motor development in breastfed infants from Porto Velho (Amazon), Brazil. *Int. J. Hyg. Environ. Health* **2007**, *210*, 51–60. [CrossRef] [PubMed]
7. Daniels, J.L.; Longnecker, M.P.; Rowland, A.S.; Golding, J.; ALSPAC Study Team. University of Bristol Institute of Child Health. Fish intake during pregnancy and early cognitive development of offspring. *Epidemiology* **2004**, *15*, 394–402. [CrossRef] [PubMed]
8. Yaginuma-Sakurai, K.; Murata, K.; Iwai-Shimada, M.; Nakai, K.; Kurokawa, N.; Tatsuta, N.; Satoh, H. Hair-to-blood ratio and biological half-life of mercury: Experimental study of methylmercury exposure through fish consumption in humans. *J. Toxicol. Sci.* **2012**, *37*, 123–130. [CrossRef] [PubMed]
9. Karagas, M.R.; Choi, A.L.; Oken, E.; Horvat, M.; Schoeny, R.; Kamai, E.; Cowell, W.; Grandjean, P.; Korrick, S. Evidence on the human health effects of low-level methylmercury exposure. *Environ. Health Perspect.* **2012**, *120*, 799–806. [CrossRef] [PubMed]
10. Grandjean, P.; Weihe, P.; White, R.F.; Debes, F.; Araki, S.; Yokoyama, K.; Murata, K.; Sørensen, N.; Dahl, R.; Jørgensen, P.J. Cognitive deficit in 7-year-old with prenatal exposure to methylmercury. *Neurotoxicol. Teratol.* **1997**, *19*, 417–428. [CrossRef]
11. Jedrychowski, W.; Jankowski, J.; Flak, E.; Skarupa, A.; Mroz, E.; Sochacka-Tatara, E.; Lisowska-Miszczyk, I.; Szpanowska-Wohn, A.; Rauh, V.; Skolicki, Z.; et al. Effects of prenatal exposure to mercury on cognitive and psychomotor function in one-year-old infants: Epidemiologic cohort study in Poland. *Ann. Epidemiol.* **2006**, *16*, 439–447. [CrossRef] [PubMed]
12. Jedrychowski, W.; Perera, F.; Jankowski, J.; Rauh, V.; Flak, E.; Caldwell, K.L.; Jones, R.L.; Pac, A.; Lisowska-Miszczyk, I. Fish consumption in pregnancy, cord blood mercury level and cognitive and psychomotor development of infants followed over the first three years of life: Krakow epidemiologic study. *Environ. Int.* **2007**, *33*, 1057–1062. [CrossRef] [PubMed]
13. Valent, F.; Mariuz, M.; Bin, M.; Little, D.; Mazej, D.; Tognin, V.; Tratnik, J.; McAfee, A.J.; Mulhern, M.S.; Parpinel, M.; et al. Associations of prenatal mercury exposure from maternal fish consumption and polyunsaturated fatty acids with child neurodevelopment: A prospective cohort study in Italy. *J. Epidemiol.* **2013**, *23*, 360–370. [CrossRef] [PubMed]

14. Hsi, H.C.; Jiang, C.B.; Yang, T.H.; Chien, L.C. The neurological effects of prenatal and postnatal mercury/methylmercury exposure on three-year-old children in Taiwan. *Chemosphere* **2014**, *100*, 71–76. [CrossRef] [PubMed]

15. Marques, R.C.; Abreu, L.; Bernardi, J.V.; Dorea, J.G. Neurodevelopment of Amazonian children exposed to ethylmercury (from Thimerosal in vaccines) and methylmercury (from fish). *Environ. Res.* **2016**, *149*, 259–265. [CrossRef] [PubMed]

16. Van Wijngaarden, E.; Thurston, S.W.; Myers, G.J.; Harrington, D.; Cory-Slechta, D.A.; Strain, J.J.; Watson, G.E.; Zareba, G.; Love, T.; Henderson, J.; et al. Methylmercury exposure and neurodevelopmental outcomes in the Seychelles Child Development Study Main cohort at age 22 and 24years. *Neurotoxicol. Teratol.* **2017**, *59*, 35–42. [CrossRef] [PubMed]

17. Prpić, I.; Milardović, A.; Vlašić-Cicvarić, I.; Špiric, Z.; Radić Nišević, J.; Vukelić, P.; Snoj Tratnik, J.; Mazej, D.; Horvat, M. Prenatal exposure to low-level methylmercury alters the child's fine motor skills at the age of 18 months. *Environ. Res.* **2017**, *152*, 369–374. [CrossRef] [PubMed]

18. Nakai, K.; Suzuki, K.; Oka, T.; Murata, K.; Sakamoto, M.; Okamura, K.; Hosokawa, T.; Sakai, T.; Nakamura, T.; Saito, Y.; et al. The Tohoku Study of Child Development: A cohort study of effects of perinatal exposures to methylmercury and environmentally persistent organic pollutants on neurobehavioral development in Japanese children. *Tohoku J. Exp. Med.* **2004**, *202*, 227–237. [CrossRef] [PubMed]

19. Suzuki, K.; Nakai, K.; Sugawara, T.; Nakamura, T.; Ohba, T.; Shimada, M.; Hosokawa, T.; Okamura, K.; Sakai, T.; Kurokawa, N.; et al. Neurobehavioral effects of prenatal exposure to methylmercury and PCBs, and seafood intake: Neonatal behavioral assessment scale results of Tohoku study of child development. *Environ. Res.* **2010**, *110*, 699–704. [CrossRef] [PubMed]

20. Iwai-Shimada, M.; Satoh, H.; Nakai, K.; Tatsuta, N.; Murata, K.; Akagi, H. Methylmercury in the breast milk of Japanese mothers and lactational exposure of their infants. *Chemosphere* **2015**, *126*, 67–72. [CrossRef] [PubMed]

21. Tatsuta, N.; Nakai, K.; Satoh, H.; Murata, K. Impacts of the Great East Japan earthquake on child's IQ. *J. Pediatr.* **2015**, *167*, 745–751. [CrossRef] [PubMed]

22. Tatsuta, N.; Murata, K.; Iwai-Shimada, M.; Yaginuma-Sakurai, K.; Satoh, H.; Nakai, K. Psychomotor ability in children prenatally exposed to methylmercury: The 18-month follow-up of Tohoku study of child development. *Tohoku J. Exp. Med.* **2017**, *242*, 1–8. [CrossRef] [PubMed]

23. Tatsuta, N.; Kurokawa, N.; Nakai, K.; Suzuki, K.; Iwai-Shimada, M.; Murata, K.; Satoh, H. Effects of intrauterine exposures to polychlorinated biphenyls, methylmercury, and lead on birth weight in Japanese male and female newborns. *Environ. Health Prev. Med.* **2017**, *22*, 39. [CrossRef] [PubMed]

24. Tatsuta, N.; Nakai, K.; Iwai-Shimada, M.; Suzuki, T.; Satoh, H.; Murata, K. Total mercury levels in hair of children aged 7 years before and after the Great East Japan Earthquake. *Sci. Total Environ.* **2017**, *596–597*, 207–211. [CrossRef] [PubMed]

25. Davidson, P.W.; Myers, G.J.; Cox, C.; Shamlaye, C.F.; Marsh, D.O.; Tanner, M.A.; Berlin, M.; Sloane-Reeves, J.; Cemichiari, E.; Choisy, O.; et al. Longitudinal neurodevelopmental study of Seychellois children following in utero exposure to methylmercury from maternal fish ingestion: Outcomes at 19 and 29 months. *Neurotoxicology* **1995**, *16*, 677–688. [PubMed]

26. Sakamoto, M.; Nakano, A.; Akagi, H. Declining Minamata male birth ratio associated with increased male fetal death due to heavy methylmercury pollution. *Environ. Res.* **2001**, *87*, 92–98. [CrossRef] [PubMed]

27. Marques, R.C.; Bernardi, J.V.; Abreu, L.; Dorea, J.G. Neurodevelopment outcomes in children exposed to organic mercury from multiple sources in a tin-ore mine environment in Brazil. *Arch. Environ. Contam. Toxicol.* **2015**, *68*, 432–441. [CrossRef] [PubMed]

28. Llop, S.; Lopez-Espinosa, M.J.; Rebagliato, M.; Ballester, F. Gender differences in the neurotoxicity of metals in children. *Toxicology* **2013**, *311*, 3–12. [CrossRef] [PubMed]

29. Gochfeld, M. Framework for gender differences in human and animal toxicology. *Environ. Res.* **2007**, *104*, 4–21. [CrossRef] [PubMed]

30. Bayley, N. *Bayley Scales of Infant Development*, 2nd ed.; Psychological Corporation: San Antonio, TX, USA, 1993.

31. Yasutake, A.; Matsumoto, M.; Yamaguchi, M.; Hachiya, N. Current hair mercury levels in Japanese: Survey in five districts. *Tohoku J. Exp. Med.* **2003**, *199*, 161–169. [CrossRef] [PubMed]

32. Koopman-Esseboom, C.; Weisglas-Kuperus, N.; de Ridder, M.A.; Van der Paauw, C.G.; Tuinstra, L.G.; Sauer, P.J. Effects of polychlorinated biphenyl/dioxin exposure and feeding type on infants' mental and psychomotor development. *Pediatrics* **1996**, *97*, 700–706. [PubMed]

33. Ministry of the Environment, Japan. Mercury Analysis Manual. 2004. Available online: http://www.nimd.go.jp/kenkyu/docs/march_mercury_analysis_manual%28e%29.pdf (accessed on 2 July 2018).

34. Nakamura, T.; Nakai, K.; Matsumura, T.; Suzuki, S.; Saito, Y.; Satoh, H. Determination of dioxins and polychlorinated biphenyls in breast milk, maternal blood and cord blood from residents of Tohoku, Japan. *Sci. Total Environ.* **2008**, *394*, 39–51. [CrossRef] [PubMed]

35. Tatsuta, N.; Suzuki, K.; Sugawara, T.; Nakai, K.; Hosokawa, T.; Satoh, H. Comparison of Kyoto Scale of Psychological Development and Bayley Scales of Infant Development second edition among Japanese infants. *J. Spec. Educ. Res.* **2013**, *2*, 17–24. [CrossRef]

36. Yaginuma-Sakurai, K.; Shimada, M.; Ohba, T.; Nakai, K.; Suzuki, K.; Kurokawa, N.; Kameo, S.; Satoh, H. Assessment of exposure to methylmercury in pregnant Japanese women by FFQ. *Public Health Nutr.* **2009**, *12*, 2352–2358. [CrossRef] [PubMed]

37. Raven, J.C. *Standard Progressive Matrices: Sets A, B, C, D and E*; Lewis: London, UK, 1985.

38. Anme, T.; Ueda, R.; Hirayama, M. Evaluation of home stimulation using HSQ (HOME screening questionnaire). *J. Child Health* **1986**, *45*, 556–560.

39. Caldwell, B.M.; Bradley, R.H. *Home Observation for Measurement of the Environment*; University of Arkansas at Little Rock: Little Rock, AK, USA, 1984.

40. Japan Food Safety Commission Secretariat. Food Safety Risk Assessment Related to Methylmercury in Seafood. 2005. Available online: http://www.fsc.go.jp/english/topics/methylmercury_risk_assessment.pdf (accessed on 2 July 2018).

41. Grandjean, P.; Budtz-Jørgensen, E.; White, R.F.; Jørgensen, P.J.; Weihe, P.; Debes, F.; Keiding, N. Methylmercury exposure biomarkers as indicators of neurotoxicity in children aged 7 years. *Am. J. Epidemiol.* **1999**, *150*, 301–305. [CrossRef] [PubMed]

42. Grandjean, P.; Weihe, P.; Debes, F.; Choi, A.L.; Budtz-Jørgensen, E. Neurotoxicity from prenatal and postnatal exposure to methylmercury. *Neurotoxicol. Teratol.* **2014**, *43*, 39–44. [CrossRef] [PubMed]

43. Davidson, P.W.; Myers, G.J.; Cox, C.; Axtell, C.; Shamlaye, C.; Sloane-Reeves, J.; Cernichiari, E.; Needham, L.; Choi, A.; Wang, Y.; et al. Effects of prenatal and postnatal methylmercury exposure from fish consumption on neurodevelopment: Outcomes at 66 months of age in the Seychelles Child Development Study. *JAMA* **1998**, *280*, 701–707. [CrossRef] [PubMed]

44. Muckle, G.; Ayotte, P.; Dewailly, E.; Jacobson, S.W.; Jacobson, J.L. Prenatal exposure of the northern Québec Inuit infants to environmental contaminants. *Environ. Health Perspect.* **2001**, *109*, 1291–1299. [PubMed]

45. Marques, R.C.; Bernardi, J.V.; Dórea, J.G.; Brandão, K.G.; Bueno, L.; Leão, R.S.; Malm, O. Fish consumption during pregnancy, mercury transfer, and birth weight along the Madeira River Basin in Amazonia. *Int. J. Environ. Res. Public Health* **2013**, *10*, 2150–2163. [CrossRef] [PubMed]

46. Cordier, S.; Garel, M.; Mandereau, L.; Morcel, H.; Doineau, P.; Gosme-Seguret, S.; Josse, D.; White, R.; Amiel-Tison, C. Neurodevelopmental investigations among methylmercury-exposed children in French Guiana. *Environ. Res.* **2002**, *89*, 1–11. [CrossRef] [PubMed]

47. Deroma, L.; Parpinel, M.; Tognin, V.; Channoufi, L.; Tratnik, J.; Horvat, M.; Valent, F.; Barbone, F. Neuropsychological assessment at school-age and prenatal low-level exposure to mercury through fish consumption in an Italian birth cohort living near a contaminated site. *Int. J. Hyg. Environ. Health* **2013**, *216*, 486–493. [CrossRef] [PubMed]

48. Ramirez, G.B.; Pagulayan, O.; Akagi, H.; Francisco Rivera, A.; Lee, L.V.; Berroya, A.; Vince Cruz, M.C.; Casintahan, D. Tagum study II: Follow-up study at two years of age after prenatal exposure to mercury. *Pediatrics* **2003**, *111*, e289-95. [CrossRef] [PubMed]

49. Gao, Y.; Yan, C.H.; Tian, Y.; Wang, Y.; Xie, H.F.; Zhou, X.; Yu, X.D.; Yu, X.G.; Tong, S.; Zhou, Q.X.; et al. Prenatal exposure to mercury and neurobehavioral development of neonates in Zhoushan City, China. *Environ. Res.* **2007**, *105*, 390–399. [CrossRef] [PubMed]

50. Gundacker, C.; Fröhlich, S.; Graf-Rohrmeister, K.; Eibenberger, B.; Jessenig, V.; Gicic, D.; Prinz, S.; Wittmann, K.J.; Zeisler, H.; Vallant, B.; et al. Perinatal lead and mercury exposure in Austria. *Sci. Total Environ.* **2010**, *408*, 5744–5749. [CrossRef] [PubMed]

51. Xue, F.; Holzman, C.; Rahbar, M.H.; Trosko, K.; Fischer, L. Maternal fish consumption, mercury levels, and risk of preterm delivery. *Environ. Health Perspect.* **2007**, *115*, 42–47. [CrossRef] [PubMed]

52. Saito, S.; Kawabata, T.; Tatsuta, N.; Kimura, F.; Miyazawa, T.; Mizuno, S.; Nishigori, H.; Arima, T.; Kagawa, Y.; Yoshimasu, K.; et al. Determinants of polyunsaturated fatty acid concentrations in erythrocytes of pregnant Japanese women from a birth cohort study: Study protocol and baseline findings of an adjunct study of the Japan Environment & Children's Study. *Environ. Health Prev. Med.* **2017**, *22*, 22. [CrossRef] [PubMed]

53. Myers, G.J.; Davidson, P.W.; Shamlaye, C.F. A review of methylmercury and child development. *Neurotoxicology* **1998**, *19*, 313–328. [PubMed]

54. McKeown-Eyssen, G.E.; Ruedy, J.; Neims, A. Methylmercury exposure in northern Quebec. II. Neurologic findings in children. *Am. J. Epidemiol.* **1983**, *118*, 470–479. [CrossRef] [PubMed]

55. Watson, G.E.; Lynch, M.; Myers, G.J.; Shamlaye, C.F.; Thurston, S.W.; Zareba, G.; Clarkson, T.W.; Davidson, P.W. Prenatal exposure to dental amalgam: Evidence from the Seychelles Child Development Study main cohort. *J. Am. Dent. Assoc.* **2011**, *142*, 1283–1294. [CrossRef] [PubMed]

56. Sørensen, N.; Murata, K.; Budtz-Jørgensen, E.; Weihe, P.; Grandjean, P. Prenatal methylmercury exposure as a cardiovascular risk factor at seven years of age. *Epidemiology* **1999**, *10*, 370–375. [CrossRef] [PubMed]

57. White, R.F.; Palumbo, C.L.; Yurgelun-Todd, D.A.; Heaton, K.J.; Weihe, P.; Debes, F.; Grandjean, P. Functional MRI approach to developmental methylmercury and polychlorinated biphenyl neurotoxicity. *Neurotoxicology* **2011**, *32*, 975–980. [CrossRef] [PubMed]

58. Sagiv, S.K.; Thurston, S.W.; Bellinger, D.C.; Amarasiriwardena, C.; Korrick, S.A. Prenatal exposure to mercury and fish consumption during pregnancy and attention-deficit/hyperactivity disorder-related behavior in children. *Arch. Pediatr. Adolesc. Med.* **2012**, *166*, 1123–1131. [CrossRef] [PubMed]

59. Watson, G.E.; Evans, K.; Thurston, S.W.; van Wijngaarden, E.; Wallace, J.M.; McSorley, E.M.; Bonham, M.P.; Mulhern, M.S.; McAfee, A.J.; Davidson, P.W.; et al. Prenatal exposure to dental amalgam in the Seychelles Child Development Nutrition Study: Associations with neurodevelopmental outcomes at 9 and 30 months. *Neurotoxicology* **2012**, *33*, 1511–1517. [CrossRef] [PubMed]

60. Llop, S.; Guxens, M.; Murcia, M.; Lertxundi, A.; Ramon, R.; Riaño, I.; Rebagliato, M.; Ibarluzea, J.; Tardon, A.; Sunyer, J.; et al. Prenatal exposure to mercury and infant neurodevelopment in a multicenter cohort in Spain: Study of potential modifiers. *Am. J. Epidemiol.* **2012**, *175*, 451–465. [CrossRef] [PubMed]

61. Olsen, J.; Sørensen, H.T. The Danish national birth cohort—A valuable tool for pharmacoepidemiology in pregnancy. *Int. J. Risk Saf. Med.* **1997**, *10*, 197–198. [CrossRef] [PubMed]

62. Magnus, P.; Irgens, L.M.; Haug, K.; Nystad, W.; Skjaerven, R.; Stoltenberg, C.; MoBa Study Group. Cohort profile: The Norwegian Mother and Child Cohort Study (MoBa). *Int. J. Epidemiol.* **2006**, *35*, 1146–1150. [CrossRef] [PubMed]

63. Duarte-Salles, T.; Mendez, M.A.; Morales, E.; Bustamante, M.; Rodríguez-Vicente, A.; Kogevinas, M.; Sunyer, J. Dietary benzo(a)pyrene and fetal growth: Effect modification by vitamin C intake and glutathione S-transferase P1 polymorphism. *Environ. Int.* **2012**, *45*, 1–8. [CrossRef] [PubMed]

64. Kawamoto, T.; Nitta, H.; Murata, K.; Toda, E.; Tsukamoto, N.; Hasegawa, M.; Yamagata, Z.; Kayama, F.; Kishi, R.; Ohya, Y.; et al. Rationale and study design of the Japan environment and children's study (JECS). *BMC Public Health* **2014**, *14*, 25. [CrossRef] [PubMed]

65. Richards, M.; Hardy, R.; Kuh, D.; Wadsworth, M.E. Birth weight and cognitive function in the British 1946 birth cohort: Longitudinal population based study. *BMJ* **2001**, *322*, 199–203. [CrossRef] [PubMed]

66. Tong, S.; Baghurst, P.; Vimpani, G.; McMichael, A. Socioeconomic position, maternal IQ, home environment, and cognitive development. *J. Pediatr.* **2007**, *151*, 284–288. [CrossRef] [PubMed]

67. Debes, F.; Budtz-Jørgensen, E.; Weihe, P.; White, R.F.; Grandjean, P. Impact of prenatal methylmercury exposure on neurobehavioral function at age 14 years. *Neurotoxicol. Teratol.* **2006**, *28*, 363–375. [CrossRef] [PubMed]

68. Debes, F.; Weihe, P.; Grandjean, P. Cognitive deficits at age 22 years associated with prenatal exposure to methylmercury. *Cortex* **2016**, *74*, 358–369. [CrossRef] [PubMed]

69. Bellinger, D.C. Interpreting epidemiologic studies of developmental neurotoxicity: Conceptual and analytic issues. *Neurotoxicol. Teratol.* **2009**, *31*, 267–274. [CrossRef] [PubMed]

70. Grandjean, P.; Weihe, P.; Burse, V.W.; Needham, L.L.; Storr-Hansen, E.; Heinzow, B.; Debes, F.; Murata, K.; Simonsen, H.; Ellefsen, P.; et al. Neurobehavioral deficits associated with PCB in 7-year-old children prenatally exposed to seafood neurotoxicants. *Neurotoxicol. Teratol.* **2001**, *23*, 305–317. [CrossRef]

71. Stewart, P.; Sargent, D.; Reihman, J.; Gump, B.; Lonky, E.; Darvill, T.; Hicks, H.; Pagano, J. Response inhibition during differential reinforcement of low rates (DRL) schedules may be sensitive to low-level polychlorinated biphenyl, methylmercury, and lead exposure in children. *Environ. Health Perspect.* **2006**, *114*, 1923–1929. [CrossRef] [PubMed]

72. Yorifuji, T.; Debes, F.; Weihe, P.; Grandjean, P. Prenatal exposure to lead and cognitive deficit in 7- and 14-year-old children in the presence of concomitant exposure to similar molar concentration of methylmercury. *Neurotoxicol. Teratol.* **2011**, *33*, 205–211. [CrossRef] [PubMed]

73. Tatsuta, N.; Nakai, K.; Murata, K.; Suzuki, K.; Iwai-Shimada, M.; Yaginuma-Sakurai, K.; Kurokawa, N.; Nakamura, T.; Hosokawa, T.; Satoh, H. Prenatal exposures to environmental chemicals and birth order as risk factors for child behavior problems. *Environ. Res.* **2012**, *114*, 47–52. [CrossRef] [PubMed]

74. Tatsuta, N.; Nakai, K.; Murata, K.; Suzuki, K.; Iwai-Shimada, M.; Kurokawa, N.; Hosokawa, T.; Satoh, H. Impacts of prenatal exposures to polychlorinated biphenyls, methylmercury and lead on intellectual ability of 42-month-old children in Japan. *Environ. Res.* **2014**, *133*, 321–326. [CrossRef] [PubMed]

75. Choi, A.L.; Mogensen, U.B.; Bjerve, K.S.; Debes, F.; Weihe, P.; Grandjean, P.; Budtz-Jørgensen, E. Negative confounding by essential fatty acids in methylmercury neurotoxicity associations. *Neurotoxicol. Teratol.* **2014**, *42*, 85–92. [CrossRef] [PubMed]

76. Strain, J.J.; Davidson, P.W.; Bonham, M.P.; Duffy, E.M.; Stokes-Riner, A.; Thurston, S.W.; Wallace, J.M.; Robson, P.J.; Shamlaye, C.F.; Georger, L.A.; et al. Associations of maternal long-chain polyunsaturated fatty acids, methylmercury and infant development in the Seychelles child development nutrition study. *Neurotoxicology* **2008**, *29*, 776–782. [CrossRef] [PubMed]

77. Kim, Y.; Ha, E.H.; Park, H.; Ha, M.; Kim, Y.; Hong, Y.C.; Lee, E.J.; Kim, H.; Chang, N.; Kim, B.N. Prenatal mercury exposure, fish intake and neurocognitive development during first three years of life: Prospective cohort mothers and Children's environmental health (MOCEH) study. *Sci. Total Environ.* **2018**, *615*, 1192–1198. [CrossRef] [PubMed]

toxics

MDPI

Article

Expression of Genes Involved in Stress, Toxicity, Inflammation, and Autoimmunity in Relation to Cadmium, Mercury, and Lead in Human Blood: A Pilot Study

Rebecca N. Monastero [1,2,*], Caterina Vacchi-Suzzi [2,3], Carmen Marsit [4], Bruce Demple [2,5] and Jaymie R. Meliker [2,6]

[1] Stony Brook University School of Medicine, 101 Nicolls Road, Health Sciences Center, Level 4, Stony Brook, NY 11794-8434, USA
[2] Stony Brook University, Stony Brook, NY 11794, USA; caterina.vacchi-suzzi@stonybrookmedicine.edu (C.V.-S.); bruce.demple@stonybrook.edu (B.D.); jaymie.meliker@stonybrook.edu (J.R.M.)
[3] Stony Brook University Cancer Center, Stony Brook Medicine 3 Edmund D. Pellegrino Road, Stony Brook, NY 11794-9452, USA
[4] Department of Environmental Health, Rollins School of Public Health, Emory University, 1518 Clifton Road, Atlanta, GA 30322, USA; carmen.j.marsit@emory.edu
[5] Department of Pharmacological Sciences, Stony Brook Medicine, BST 8-140, Stony Brook, NY 11794, USA
[6] Program in Public Health, Department of Family, Population and Preventive Medicine, Stony Brook University, HSC L3, Rm 071, Stony Brook, NY 11794, USA
* Correspondence: rnmonastero@gmail.com; Tel.: +1-631-444-1145

Received: 11 June 2018; Accepted: 4 July 2018; Published: 6 July 2018

Abstract: There is growing evidence of immunotoxicity related to exposure to toxic trace metals, and an examination of gene expression patterns in peripheral blood samples may provide insights into the potential development of these outcomes. This pilot study aimed to correlate the blood levels of three heavy metals (mercury, cadmium, and lead) with differences in gene expression in 24 participants from the Long Island Study of Seafood Consumption. We measured the peripheral blood mRNA expression of 98 genes that are implicated in stress, toxicity, inflammation, and autoimmunity. We fit multiple linear regression models with multiple testing correction to correlate exposure biomarkers with mRNA abundance. The mean blood Hg in this cohort was 16.1 µg/L, which was nearly three times the Environmental Protection Agency (EPA) reference dose (5.8 µg/L). The levels of the other metals were consistent with those in the general population: the mean Pb was 26.8 µg/L, and the mean Cd was 0.43 µg/L. The expression of three genes was associated with mercury, four were associated with cadmium, and five were associated with lead, although none were significant after multiple testing correction. Little evidence was found to associate metal exposure with mRNA abundance for the tested genes that were associated with stress, toxicity, inflammation, or autoimmunity. Future work should provide a more complete picture of physiological reactions to heavy metal exposure.

Keywords: Hg; Cd; Pb; mRNA; Fish

1. Introduction

Three toxic metals to which humans are commonly exposed include mercury, cadmium, and lead [1]. Hg, Cd, and Pb are known to influence many diseases and conditions, including but not limited to cancer and neurologic, renal, and bone diseases [2–4]. Exposure to these metals may affect human health due to varied mechanisms of action, including their possible alteration of gene expression [5,6]. Despite most of the population having only moderate exposure to these metals, which is the equivalent of approximately 1 µg/kg/day for Pb, 0.30–0.35 µg/kg/day for Cd, and approximately 0.12 µg/kg/day for total Hg [2–4,7–9], it is important to characterize how the human body reacts upon exposure. At levels of exposure that are common in the general population, the associations of some of these metals with disease states including coronary heart disease, kidney disease, various autoimmune diseases, neurological and neurodegenerative disorders, and myocardial infarction have been observed [10–12]. Specifically, methylmercury (MeHg) exposure has been most strongly associated with neurotoxicity, including neurodevelopmental delays in children and decreased manual dexterity and fine motor speed in adults, as well as impaired verbal skills, memory, and visual motor functions in both children and adults. Cd has been associated with kidney disease in industrially-exposed populations, and gastrointestinal complications and bone fragility in those with low-level oral exposure. Pb has been associated with neuropathy, reproductive toxicity, renal disease, and hypertension in adults, and loss of IQ in children [2,3,13]. Importantly, over the past century, exposure to these heavy metals has been on the rise in the developing population, resulting from increases in heavy metal use in agriculture, technology, and industry [14]. The frequent exposures to these metals and their health implications emphasize the necessity for evaluating the risks of these toxicants not only clinically, but at the molecular level as well, for a better understanding of the human body's biochemical reactions to these environmental pollutants and providing potential early biomarkers of exposure.

These heavy metals may induce their systemic effects through changes in gene expression, as the abundance of gene transcripts (mRNA) is often altered in toxic, immune, and autoimmune responses. A limited number of in vitro studies reported correlations between metal exposure and mRNAs encoding genes related to stress, toxicity, and autoimmunity/immunity. Kawata et al. (2007) investigated alterations in gene expression in human HepG2 cells upon heavy metal exposure compared to known chemical responses, and they found that large numbers of the chemically-inducible genes were also responsive to Cd and Hg [15]. In human fibroblasts, Li et al. (2008) identified 35 genes that were responsive to Cd, most of which were associated with cell cycle, immunity/defense, nucleoside metabolism, and signal transduction [16]. Lee et al. (2013) also investigated altered RNA expression due to Cd, and they found the upregulated expression of 30 genes in HK-2 human proximal tubular cells, including genes that were involved in transcription and heat shock [17].

Animal studies, specifically murine models, have provided additional insight into the physiological response to heavy metals, although these efforts have also been limited in number and primarily investigated the effects of Cd. Murine models of metal exposure and gene expression in various organs (including intestine, kidney, liver, and testes) have demonstrated Cd-modulated and Hg-modulated expression of genes encoding functions related to transport, oxidative stress, and inflammation: the modulation genes related to heat shock, acute phase, metallothionein, and antioxidants due to Cd and Hg, and the modulation of genes related to angiogenesis, hypoxia, toxicity, and carcinogenesis due to Cd [18–21]. However, these animal studies primarily involved high-dose exposures to heavy metals measured in organs, which must be considered when comparing results to lower-level human dietary or occupational metal exposures assessed in peripheral blood.

There is currently a dearth of epidemiologic studies investigating how heavy metals correlate with gene expression in humans at more population-relevant exposure levels. Changes in expression are often linked to disease initiation and progression, and environmental exposures to toxins can modulate this genetic predisposition in the human system [22]. A limited number of epidemiologic studies have noted the effect of long-term Hg, Cd, or Pb exposure, as measured in blood and urine, on the modulated expression in blood of mRNA involved in a variety of oxidative stress, toxicity, and DNA repair pathways, suggesting an effect of these metals on the human body's response to stress [23,24]. For experimental short-term, gaseous metal fume exposure, Wang et al. (2005) similarly observed decreases in oxidative stress response transcripts in blood, in addition to decreases in apoptosis-related and inflammatory response-related transcripts [25]. A recent study by Korashy et al. (2017) also noted the modulation of mRNA abundance in blood in response to blood Hg, Cd, or Pb, in this case with a large number of genes upregulated, and a smaller number of genes downregulated in response to the metal exposure [26]. Despite these limited studies suggesting varied stress responses correlated with metal exposure, studies have not looked at transcript abundance across a number of genes, necessitating work in this area to better understand how physiological pathways are modulated in response to metal exposure, rather than only individual genes. Furthermore, the correlations reported to date are not consistent for genes in related pathways, and merit additional scrutiny.

In this pilot study, we investigated gene expression associations with metal exposure in humans, testing the hypothesis that increased peripheral blood levels of total Hg, Cd, and Pb are correlated with increased expression of 98 genes known to be involved with response processes for inflammation, autoimmunity, oxidative stress, and toxicity.

2. Methods

2.1. Participant Recruitment

The study was approved by Stony Brook University's Committee on Research Involving Human Subjects (CORIHS) (IRB #2010-1179). Between 2011–2012, 290 avid seafood consumers were recruited from the general Long Island population. Participants were recruited in-person and through flyers posted at fishing piers, seafood restaurants, markets, gyms, libraries, university bulletin boards, and three advertisements and an article about the study that ran in a newspaper (Newsday).

Interested participants (996 individuals) were informed that the study was being conducted to evaluate the risks and benefits of seafood consumption. Interested participants filled out surveys of self-reported seafood consumption to evaluate their approximate expected blood Hg levels based on seafood Hg concentrations reported in Karimi et al. (2012) [27]. The estimated blood Hg cutoff that was used to determine eligibility for this study to ensure adequate power was an estimated whole blood concentration of 5.8 µg/L, which corresponds to the United States (US) EPA reference dose of 0.1 µg/kg/day. This cutoff yielded 746 eligible participants, of which 290 enrolled in our study. Eligible participants completed questionnaires to collect demographics, among other factors.

For this study, 24 individuals were selected for analysis of both metals and mRNA expression. These individuals were selected because they were among those with the highest and lowest levels of each metal type, thereby ensuring a range of exposure levels for each metal. Age, gender, current smoking status, and levels of omega-3 fatty acids in plasma (for laboratory methods, see Karimi et al. 2014) were extracted from the database for use in regression analyses [28].

2.2. Blood Biomarkers: Collection and Analysis of Blood Hg, Cd, and Pb

Blood biomarkers were collected and used for analysis due to their low variability, long half-life, and reflection of body burden resulting from representation of organ metal uptake [29]. Whole blood was collected by venipuncture for the measurement of heavy metals (Hg, Cd, and Pb) in trace metal tubes containing K2EDTA. Blood specimens were stored at 4 °C and sent to RTI International's Trace Inorganics Laboratory (Research Triangle Park, NC, USA) for analysis of Hg, Cd, and Pb by ICP-MS (Thermo-X Series II). A 1000 µg/mL gold solution (High Purity Standards) was added to samples for heavy metal stability. Samples were microwave-digested with nitric acid and hydrogen peroxide (J.T. Baker, Ultrex Grade), and diluted with deionized water. Standard reference materials, including bovine and caprine blood (NIST SRM955c caprine blood, NIST SRM966 bovine blood, and UTAK human blood), were also digested and analyzed for quality control, with an average recovery percentage of $110 \pm 14\%$ ($n = 9$). Sample blanks and method blanks showed negligible Hg, Cd, and Pb, confirming a lack of background contamination due to the sample collection method and digestion method, respectively. Three samples were found to be below the detection limit (LOD) (0.3 µg/L) for Hg, and all of the samples were above the LOD for Pb (0.1 µg/L) and for Cd (0.04 µg/L).

2.3. Gene Expression by Reverse Transcription-Polymerase Chain Reaction PCR (RT-PCR)

Whole blood was collected via venipuncture from 13 females and 11 males ($n = 24$) for analysis of gene expression. Participants were selected to ensure a range of blood metal levels that was consistent with the entire study population. Whole blood samples were stored in PAXgene Blood RNA tubes at $-80\ ^{\circ}$C (BD Diagnostics, Franklin Lakes, NJ, USA), following instructions from the manufacturer. Samples were thawed at room temperature before further processing for at least two hours to ensure the digestion of cellular debris and stability of RNA. The PAXgene Blood miRNA kit (Qiagen, Valencia, CA, USA) was used to isolate total RNA in 80 µL of supplied elution buffer. Mean RNA Integrity Number (RIN) was 7.91 (min 7.0, max 8.7), as measured via a Bioanalyzer RNA chip (Agilent, Santa Clara, CA, USA). The average ratio of absorbance at 260 nm and 280 nm was 2.04 (min 1.94, max 2.13), which was measured via Nanodrop 2000 (Nanodrop, Wilmington, DE, USA).

Isolated RNA (300 ng) was reverse transcribed to cDNA using an RT^2 First Strand Kit (Qiagen, CA), as per the manufacturer's instruction. For each sample, 102 µL of cDNA were loaded onto RT^2 Profiler PCR Array Human Stress & Toxicity (PAHS-003ZA) and Human Inflammatory Response and Autoimmunity (PAHS-077ZA) 96-well plates, along with 1248 µL of RNAse-free water and 1350 µL of 2X SYBR Green ROX qPCR Mastermix (Qiagen, CA; Tables S1 and S2). Plates were sealed using optical adhesive film and run on an Applied Biosystems 7300 Real-Time PCR System using standard run settings.

Ct detection threshold was set at 0.2 for each sample to compare results across the dataset. Ct values were then extracted for each gene and normalized against the geometric mean of β-actin and GAPDH housekeeping (HK) genes to obtain ΔCt values. ΔCt values were then unlogged ($2^{-\Delta Ct}$), obtaining a value proportional to the relative abundance level in the sample. Probes with Ct values above 30 (set LOD) were excluded, and any gene with one or more missing values across the sample's set was excluded. Out of the 168 measured genes, 98 genes were consistently detected in the blood of all 24 test subjects, and were carried forward in the analysis.

2.4. Data and Statistical Analysis

Linear regression models were constructed to investigate the correlations between gene expression and each measured metal (Hg, Cd, Pb), with each gene as an individual dependent variable. All of the models were adjusted by age and gender. Metal levels were not associated with smoking status or with omega-3 fatty acids; therefore, in order to maintain parsimony given the small sample size, these variables were not included in the model. Adjusted R square, beta coefficient, *p*-value, and corrected *p*-value were reported to be significantly associated with one or more metals for each gene. To address multiple testing, the false discovery rate *q*-value (Benjamini–Hochberg) was calculated, and is reported for those genes that were significantly associated with a metal prior to multiple testing correction [30].

3. Results

Control ("housekeeping") genes represented in the arrays were detected at similar levels across the individual plates. Probes did not detect matching DNA strands (HGDC), indicating effective RNA isolation. The average blood Hg in participants was 16.1 µg/L, which was approximately three times higher than the RfD. Average blood Cd and Pb were 0.42 µg/L and 26.8 µg/L, respectively (Table 1). There was no significant difference between males and females for the measured markers.

Table 1. Characteristics of the study cohort ($n = 24$), 13 females, 11 males.

Variable	Average	SD	Min.	Max.	T Test Average Difference *p* by Sex
Age (years)	58	13	19	78	0.82
Hg (µg/L)	16.1	14.5	0.30	44.6	0.32
Pb (µg/L)	26.8	12.6	11.4	60.1	0.29
Cd (µg/L)	0.43	0.36	0.01	1.47	0.85

Three genes were significantly associated ($p < 0.05$) with Hg, four were significantly associated with Cd, and five were significantly associated with lead (Table 2). Without multiple testing correction, Hg was significantly associated with *IL1RAP*, *CXCR1*, and *ITGB2*, Cd was significantly associated with *ITGB2*, *C3AR1*, *TLR9*, and *TNFRSF10A*, and Pb was significantly associated with *TNFRSF1A*, *VEGFA*, *MMP9*, *ULK1*, and *SQSTM1*. However, no genes were significantly correlated with metals after accounting for multiple testing.

Table 2. Whole blood expression of genes associated with blood Hg, Cd, or Pb levels (non-adjusted $p < 0.05$).

Gene Symbol	Associated Metal	Adjusted R Square	B	*p* Value	BH-Corrected p
IL1RAP	Hg	0.36	−0.34	0.03	0.95
CXCR1	Hg	0.19	−2.26	0.04	0.95
ITGB2	Hg	0.11	−1.95	0.05	0.95
ITGB2	Cd	0.18	89.13	0.02	0.87
C3AR1	Cd	0.34	14.22	0.01	0.87
TLR9	Cd	0.16	−1.61	0.04	0.96
TNFRSF10A	Cd	0.12	−2.75	0.04	0.96
VEGFA	Pb	0.37	0.04	0.01	0.50
TNFRSF1A	Pb	0.19	1.29	0.01	0.50
ULK1	Pb	0.22	0.14	0.02	0.50
SQSTM1	Pb	0.14	1.41	0.02	0.50
MMP9	Pb	0.11	1.32	0.03	0.59

All models adjusted by age and sex.

4. Discussion

After multiple testing correction, our results show no significant associations between peripheral blood Hg, Cd, and Pb and mRNA levels that are indicative of stress, toxicity, immune response, or autoimmunity. However, given that this is a small pilot study, the significant associations that are present between these metals and certain genes without adjustment for multiple testing may help direct a searchlight for future studies with larger sample sizes and potentially higher exposures to heavy metals. Therefore, even though the results were not significant after multiple testing correction, we feel that it is important to consider the biologic plausibility of those genes that were associated prior to multiple testing correction.

The expression of each of the three genes that were associated with total blood Hg prior to multiple testing correction—*IL1RAP*, *CXCR1*, and *ITGB2*—decreased as Hg levels increased. These three genes are markers for inflammatory responses, as indicated by the Inflammatory Response and Autoimmunity PCR Assay we employed, suggesting that the inflammatory response induced by these three gene products may be dampened at higher levels of Hg. In a sensitivity analysis, we repeated the multiple regression analyses additionally adjusting for omega-3s as a percentage of fatty acids in phospholipid fraction of plasma, and for the self-reported frequency of total fish intake. Beta coefficients and directions of association were consistent, suggesting that confounding by fish intake is not influencing our findings. However, biologic plausibility must be considered, as in previous work, relatively toxic doses of heavy metals were required to achieve significant changes in gene expression in animal models, suggesting that higher exposure levels would provide a more considerable effect upon expression of these genes in humans as well [19]. Nonetheless, all three gene products are known to be involved in neutrophil-dependent pathways and pathways downstream of neutrophil activation, as they are activated largely through NF-κB pathways, and subsequently activate effector pathways or provide negative feedback to neutrophils. This suggests that their similar associations with Hg in this study may be due to the gene products' functional relationships in stimulation of immune actors. IL-8, which is a ligand for the gene product of *CXCR1*, and the integrin β-2, are both known to be involved in stimulating neutrophil activation and migration, which in turn stimulates the release of a variety of cytokines and chemokines through an NF-κB pathway in both murine and human models [31–34]. Cytokines released by neutrophils include IL-1β and IL-1α; these are ligands for the receptor that is associated with the gene product of *IL-1RAP*, which, similar to IL-8, also serve to both stimulate neutrophil activation and migration [34–36]. Therefore, the gene products of our Hg-associated genes play a role in regulation of the inflammatory response through release by NF-κB pathways or the promotion of those pathways, suggesting a common modulation by Hg that may become more apparent in a study with a larger sample size.

C3AR1, similar to ITGB2, was positively associated with Cd, whereas *TLR9* and *TNFRSF10A* were negatively associated with Cd prior to multiple testing correction. *MMP9*, *VEGFA*, *ULK1*, *SQSTM1*, and *TNFRSF1A* were all positively associated with Pb prior to multiple testing correction. Studies have not been conducted epidemiologically nor at the level of toxicity to evaluate the modulation of expression of these genes by these metals, and therefore, no data-driven hypotheses can be developed. Dose-response curves have not been established for Cd with respect to gene expression, and therefore, biologic plausibility is unable to be evaluated primarily due to a lack of data at relevant doses.

One challenge in this area of study that impacts the interpretation of the metal-associated genes is the dual immunosuppressive and immunostimulatory effects of heavy metals, which are currently not well-characterized [5]. Unlike some immunotoxins, both effects due to metals have been found in animal and human studies, highlighting the difficult task of identifying which pathways in particular are affected and interacting. Metals as immunotoxins do not act in immunosuppressive ways specifically, but also may play a role in stimulating allergic responses, autoimmunity, or infection susceptibility [8]. This differential effect of metals on humans characterizes metals more accurately as immunomodulators rather than immunotoxins, and suggests that future work be focused more on distinguishing these effects at different doses. Such pathway modulations must be studied in more depth through studies of entire pathways, rather than individual protein products; i.e. examining multiple protein products in a pathway, in order to better understand exactly which products are being upregulated and downregulated to result in an outcome pathway modulation.

A limitation that should be considered is the potential for metal toxicity to be affected by a variety of factors that can alter how inflammatory/stress/autoimmune/toxicity markers are modulated. Factors affecting metal toxicity include the form of the metal, level of exposure, duration of exposure, mode of exposure, and toxicokinetics/toxicodynamics of the compound [6]. We measured total Hg in the blood among a population of seafood consumers; therefore, this is likely to reflect methylHg exposure, which is the dominant form of Hg in fish. Certain populations are also more susceptible to metal toxicity, including those with glutathione deficiencies, thalassemia, or cystinuria, and adolescents or older individuals; however, we were able to control only for age in this study. Blood measures of metals may be associated with potential confounding variables, such as cigarette smoking, although we did not observe associations between smoking and metal levels in this study. Nonetheless, the possibility of a confounding factor influencing the metal levels and gene expression cannot be ruled out. Gene expression can also appear to be modulated, particularly in one-time mRNA measurements, due to a lack of stability of mRNA transcripts over time. The half-lives of mRNA transcripts vary widely, and can depend on a number of extracellular signals and factors that regulate transcript degradation at different levels for the modulation of gene expression [37]. Due to such extensive variation in mRNA half-life and stability, one-time measurements of mRNA are not ideal for representing gene expression. Future work may look at particular metal compounds at varying levels, durations, and routes of exposure in a wider variety of individuals to obtain a more complete picture of their effects on particular gene markers measured over time.

Another important limitation of this study involves the difficulty of evaluating graded transcript abundance as a measure of gene upregulation or downregulation. The upregulation and downregulation of genes may be evident in only a small transcript abundance modulation, whereas the modulation of other genes may present as large a difference in transcript abundance. To account for this variation in expression response, our analysis focused only on positive or negative changes, sans gradient. Future work may seek to examine graded transcript abundance modulations for an improved understanding of dosage-dependence and relative contributions/interactions of gene products in the discussed pathways.

To conclude, in this small pilot study of 24 individuals, we report no association between Hg, Cd, or Pb and any of 98 mRNA markers of stress, toxicity, inflammation, or autoimmunity in seafood consumers with elevated blood Hg. However, the associations that were present prior to multiple testing adjustment may point toward genetic pathways for future work in larger populations with multiple measurements across time in order to better understand how the human body responds to heavy metal exposure.

Supplementary Materials: The following are available online at http://www.mdpi.com/2305-6304/6/3/35/s1, Table S1: SaBiosciences PCR Array Inflammatory Response and Autoimmunity Genes with Detectable Data for All 24 Participants (PAHS-077Z), Table S2: SaBiosciences PCR Array Human Stress and Toxicity Genes with Detectable Data for All 24 Participants (PAHS-003Z).

Author Contributions: Conceptualization, C.M., B.D. and J.R.M.; Formal analysis, R.N.M. and C.V.-S.; Funding acquisition, C.M., B.D. and J.R.M.; Methodology, C.V.-S. and B.D.; Project administration, C.V.-S.; Supervision, B.D. and J.R.M.; Writing—original draft, R.N.M.; Writing—review & editing, C.V.-S., C.M., B.D. and J.R.M.

Funding: This work was supported by the Gelfond Fund for Mercury Research and Outreach under NIEHS R01ES019209-03S1. Funder had no involvement in this research.

Acknowledgments: This work was generously supported by the Gelfond Fund for Mercury Research and Outreach and NIEHS R01ES019209-03S1; we also thank the participants for their time and interest in our study.

Conflicts of Interest: Authors disclose they have no conflicts of interest, financial, personal, or otherwise.

Human Subjects: Our study was conducted in accordance with the Declaration of Helsinki and Stony Brook University's Committee on Research Involving Human Research Subjects (CORIHS) approved the study (IRB #2010-1179, 2 December 2010). Signed informed consent was provided. The work described in the article was carried out in accordance with The Code of Ethics of the World Medical Association, and was carried out and written in accordance with the Uniform Requirements for Manuscripts Submitted to Biomedical Journals.

References

1. ATSDR. The 2011 ATSDR Substance Priority List. 2011. Available online: https://www.atsdr.cdc.gov/spl/resources/2011_atsdr_substance_priority_list.html (accessed on 25 March 2017).

2. ATSDR. Toxicological Profile for Lead cas#7439-92-1. 2007. Available online: https://www.atsdr.cdc.gov/toxprofiles/TP.asp?id=96&tid=22 (accessed on 25 March 2017).

3. ATSDR. Toxicological Profile for Cadmium cas#7440-43-9. 2012. Available online: https://www.atsdr.cdc.gov/toxprofiles/tp.asp?id=48&tid=15 (accessed on 25 March 2017).

4. ATSDR. Toxicological Profile for Mercury cas#7439-97-6. 1999. Available online: https://www.atsdr.cdc.gov/toxprofiles/tp.asp?id=115&tid=24 (accessed on 25 March 2017).

5. Sweet, L.I.; Zelikoff, J.T. Toxicology and immunotoxicology of mercury: A comparative review in fish and humans. *J. Toxicol. Environ. Health Part B Crit. Rev.* **2001**, *4*, 161–205. [CrossRef]

6. Kakkar, P.; Jaffery, F.N. Biological markers for metal toxicity. *Environ. Toxicol. Pharmacol.* **2005**, *19*, 335–349. [CrossRef] [PubMed]

7. Osuna, C.E.; Grandjean, P.; Weihe, P.; El-Fawal, H.A. Autoantibodies associated with prenatal and childhood exposure to environmental chemicals in faroese children. *Toxicol. Sci.* **2014**, *142*, 158–166. [CrossRef] [PubMed]

8. Carey, J.B.; Allshire, A.; van Pelt, F.N. Immune modulation by cadmium and lead in the acute reporter antigen-popliteal lymph node assay. *Toxicol. Sci.* **2006**, *91*, 113–122. [CrossRef] [PubMed]

9. Coelho, P.; Garcia-Leston, J.; Costa, S.; Costa, C.; Silva, S.; Fuchs, D.; Geisler, S.; Dall'Armi, V.; Zoffoli, R.; Bonassi, S.; et al. Immunological alterations in individuals exposed to metal(loid)s in the Panasqueira mining area, central Portugal. *Sci. Total Environ.* **2014**, *475*, 1–7. [CrossRef] [PubMed]

10. Houston, M.C. The role of mercury and cadmium heavy metals in vascular disease, hypertension, coronary heart disease, and myocardial infarction. *Altern. Ther. Health Med.* **2007**, *13*, S128–S133. [PubMed]

11. Hodgson, S.; Nieuwenhuijsen, M.J.; Elliott, P.; Jarup, L. Kidney disease mortality and environmental exposure to mercury. *Am. J. Epidemiol.* **2007**, *165*, 72–77. [CrossRef] [PubMed]
12. Pollard, K.M.; Hultman, P.; Kono, D.H. Toxicology of autoimmune diseases. *Chem. Res. Toxicol.* **2010**, *23*, 455–466. [CrossRef] [PubMed]
13. Karagas, M.R.; Choi, A.L.; Oken, E.; Horvat, M.; Schoeny, R.; Kamai, E.; Cowell, W.; Grandjean, P.; Korrick, S. Evidence on the human health effects of low-level methylmercury exposure. *Environ. Health Perspect.* **2012**, *120*, 799–806. [CrossRef] [PubMed]
14. Bradl, H. *Heavy Metals in the Environment: Origin, Interaction, and Remediation*; Academic Press: London, UK, 2002; Volume 6, pp. 17–27.
15. Kawata, K.; Yokoo, H.; Shimazaki, R.; Okabe, S. Classification of heavy-metal toxicity by human DNA microarray analysis. *Environ. Sci. Technol.* **2007**, *41*, 3769–3774. [CrossRef] [PubMed]
16. Li, G.Y.; Kim, M.; Kim, J.H.; Lee, M.O.; Chung, J.H.; Lee, B.H. Gene expression profiling in human lung fibroblast following cadmium exposure. *Food Chem. Toxicol.* **2008**, *46*, 1131–1137. [CrossRef] [PubMed]
17. Lee, J.Y.; Tokumoto, M.; Fujiwara, Y.; Satoh, M. Gene expression analysis using DNA microarray in hk-2 human proximal tubular cells treated with cadmium. *J. Toxicol. Sci.* **2013**, *38*, 959–962. [CrossRef] [PubMed]
18. Breton, J.; Le Clere, K.; Daniel, C.; Sauty, M.; Nakab, L.; Chassat, T.; Dewulf, J.; Penet, S.; Carnoy, C.; Thomas, P.; et al. Chronic ingestion of cadmium and lead alters the bioavailability of essential and heavy metals, gene expression pathways and genotoxicity in mouse intestine. *Arch. Toxicol.* **2013**, *87*, 1787–1795. [CrossRef] [PubMed]
19. Bartosiewicz, M.J.; Jenkins, D.; Penn, S.; Emery, J.; Buckpitt, A. Unique gene expression patterns in liver and kidney associated with exposure to chemical toxicants. *J. Pharmacol. Exp. Ther.* **2001**, *297*, 895–905. [PubMed]
20. Eyssen-Hernandez, R.; Ladoux, A.; Frelin, C. Differential regulation of cardiac heme oxygenase-1 and vascular endothelial growth factor mrna expressions by hemin, heavy metals, heat shock and anoxia. *FEBS Lett.* **1996**, *382*, 229–233. [CrossRef]
21. Zhou, T.; Jia, X.; Chapin, R.E.; Maronpot, R.R.; Harris, M.W.; Liu, J.; Waalkes, M.P.; Eddy, E.M. Cadmium at a non-toxic dose alters gene expression in mouse testes. *Toxicol. Lett.* **2004**, *154*, 191–200. [CrossRef] [PubMed]
22. Edwards, T.M.; Myers, J.P. Environmental exposures and gene regulation in disease etiology. *Cien Saude Colet* **2008**, *13*, 269–281. [CrossRef] [PubMed]
23. Al Bakheet, S.A.; Attafi, I.M.; Maayah, Z.H.; Abd-Allah, A.R.; Asiri, Y.A.; Korashy, H.M. Effect of long-term human exposure to environmental heavy metals on the expression of detoxification and DNA repair genes. *Environ. Pollut.* **2013**, *181*, 226–232. [CrossRef] [PubMed]
24. Pizzino, G.; Bitto, A.; Interdonato, M.; Galfo, F.; Irrera, N.; Mecchio, A.; Pallio, G.; Ramistella, V.; De Luca, F.; Minutoli, L.; et al. Oxidative stress and DNA repair and detoxification gene expression in adolescents exposed to heavy metals living in the milazzo-valle del mela area (Sicily, Italy). *Redox Biol.* **2014**, *2*, 686–693. [CrossRef] [PubMed]
25. Wang, Z.; Neuburg, D.; Li, C.; Su, L.; Kim, J.Y.; Chen, J.C.; Christiani, D.C. Global gene expression profiling in whole-blood samples from individuals exposed to metal fumes. *Environ. Health Perspect.* **2005**, *113*, 233–241. [CrossRef] [PubMed]
26. Korashy, H.M.; Attafi, I.M.; Famulski, K.S.; Bakheet, S.A.; Hafez, M.M.; Alsaad, A.M.S.; Al-Ghadeer, A.R.M. Gene expression profiling to identify the toxicities and potentially relevant human disease outcomes associated with environmental heavy metal exposure. *Environ. Pollut.* **2017**, *221*, 64–74. [CrossRef] [PubMed]
27. Karimi, R.; Fitzgerald, T.P.; Fisher, N.S. A quantitative synthesis of mercury in commercial seafood and implications for exposure in the united states. *Environ. Health Perspect.* **2012**, *120*, 1512–1519. [CrossRef] [PubMed]
28. Karimi, R.; Fisher, N.S.; Meliker, J.R. Mercury-nutrient signatures in seafood and in the blood of avid seafood consumers. *Sci. Total Environ.* **2014**, *496*, 636–643. [CrossRef] [PubMed]
29. Barbosa, F., Jr.; Tanus-Santos, J.E.; Gerlach, R.F.; Parsons, P.J. A critical review of biomarkers used for monitoring human exposure to lead: Advantages, limitations, and future needs. *Environ. Health Perspect.* **2005**, *113*, 1669–1674. [CrossRef] [PubMed]
30. Benjamini, Y.; Hochberg, Y. Controlling the false discovery rate: A practical and powerful approach to multiple testing. *J. R. Stat. Soc. Ser. B* **1995**, *57*, 289–300.

31. Awla, D.; Abdulla, A.; Zhang, S.; Roller, J.; Menger, M.D.; Regner, S.; Thorlacius, H. Lymphocyte function antigen-1 regulates neutrophil recruitment and tissue damage in acute pancreatitis. *Br. J. Pharmacol.* **2011**, *163*, 413–423. [CrossRef] [PubMed]
32. McDonald, P.P.; Bald, A.; Cassatella, M.A. Activation of the nf-kappab pathway by inflammatory stimuli in human neutrophils. *Blood* **1997**, *89*, 3421–3433. [PubMed]
33. Roebuck, K.A. Regulation of interleukin-8 gene expression. *J. Interf. Cytokine Res.* **1999**, *19*, 429–438. [CrossRef] [PubMed]
34. Tecchio, C.; Micheletti, A.; Cassatella, M.A. Neutrophil-derived cytokines: Facts beyond expression. *Front. Immunol.* **2014**, *5*, 508. [CrossRef] [PubMed]
35. Oliveira, S.H.; Canetti, C.; Ribeiro, R.A.; Cunha, F.Q. Neutrophil migration induced by IL-1beta depends upon LTB4 released by macrophages and upon TNF-alpha and IL-1beta released by mast cells. *Inflammation* **2008**, *31*, 36–46. [CrossRef] [PubMed]
36. Perretti, M.; Flower, R.J. Modulation of IL-1-induced neutrophil migration by dexamethasone and lipocortin 1. *J. Immunol.* **1993**, *150*, 992–999. [PubMed]
37. Chen, C.Y.; Ezzeddine, N.; Shyu, A.B. Messenger RNA half-life measurements in mammalian cells. *Methods Enzymol.* **2008**, *448*, 335–357. [PubMed]

MDPI

St. Alban-Anlage 66

4052 Basel

Switzerland

Tel. +41 61 683 77 34

Fax +41 61 302 89 18

www.mdpi.com

Toxics Editorial Office

E-mail: toxics@mdpi.com

www.mdpi.com/journal/toxics

www.ingramcontent.com/pod-product-compliance
Lightning Source LLC
Chambersburg PA
CBHW051909210326
41597CB00033B/6077